The Three Failures
of Creationism

The Three Failures
of Creationism

Logic, Rhetoric, and Science

Walter M. Fitch

UNIVERSITY OF CALIFORNIA PRESS

Berkeley   Los Angeles   London

University of California Press, one of the most distinguished university presses in the United States, enriches lives around the world by advancing scholarship in the humanities, social sciences, and natural sciences. Its activities are supported by the UC Press Foundation and by philanthropic contributions from individuals and institutions. For more information, visit www.ucpress.edu.

University of California Press
Berkeley and Los Angeles, California

University of California Press, Ltd.
London, England

Library of Congress Cataloging-in-Publication Data

Fitch, Walter M., 1929–2011.
    The three failures of creationism: logic, rhetoric, and science / Walter M. Fitch.
        p. cm.
    Includes bibliographical references and index.
    ISBN 978-0-520-27053-4 (pbk. : alk. paper)
    1. Creationism.   2. Religion and science.   3. Logic.   4. Evolution (Biology)
5. Faith and reason   I. Title.
    BS651.F54   2012
    231.7′652—dc23   2011018982

Manufactured in the United States of America

21   20   19   18   17   16   15   14   13   12
10   9   8   7   6   5   4   3   2   1

In keeping with a commitment to support environmentally responsible and sustainable printing practices, UC Press has printed this book on 50-pound Enterprise, a 30% post-consumer-waste, recycled, deinked fiber that is processed chlorine-free. It is acid-free and meets all ANSI/NISO (z 39.48) requirements.

To my soul mate Chung Cha

CONTENTS

FOREWORD

The theory of biological evolution is the central organizing principle of modern biology. In 1973, the eminent evolutionist Theodosius Dobzhansky famously asserted, "Nothing in biology makes sense except in the light of evolution." Evolution provides a scientific explanation for why there are so many different kinds of organisms on Earth and gives an account of their similarities and differences (morphological, physiological, and genetic). It accounts for the appearance of humans on Earth and reveals our species's biological connections with other living things. It provides an understanding of the constantly evolving bacteria and viruses and other pathogenic organisms, and it enables the development of effective new ways to protect ourselves against the diseases they cause. Knowledge of evolution has made possible improvements in agriculture and medicine, and has been applied in many fields outside biology—for example, in software engineering, where genetic algorithms seek to mimic selective processes; and in chemistry, where the principles of natural selection are used for developing new molecules with specific functions.

Yet, in the United States, many people reject the scientific knowledge concerning evolution, mostly for religious reasons. According to a Gallup poll of 1,016 U.S. adults, taken in November 2004, 45 percent of those surveyed favored the statement "God created human beings in their present form within the last 10,000 years." Thirty-eight percent favored "Man developed over millions of years, but God guided the process," and 13 percent favored "Man developed over millions of years from less advanced life forms." Teaching creationism rather than evolution in the schools is favored by a large number of American citizens. In a CNN/*USA Today* Gallup poll of 1,001 adults conducted in March 2005, 76 percent would not "be upset if public schools in [their] community taught creationism," but only 63 percent would not "be upset if the schools taught evolution." Only 22 percent would be upset if creationism was taught, while 34 percent would be upset if evolution was taught. Other polls yield similar statistics.

Are evolution and religion—or, more generally, science and religion—in contradiction? No. In fact, if they are properly understood, science and religion cannot be in contradiction, because science and religion concern different aspects of the human experience. Science and religion are like two different windows for looking at the world. Both look at the same world, but they show different aspects of that world. Science encompasses the processes that account for the natural world: how planets move, the composition of matter and the atmosphere, the origin and adaptations of organisms, and so on. Religion concerns the meaning and purpose of the world and of human life, the proper relation of people to the creator and to each other, the moral vales that inspire and govern people's lives, and more. Apparent contradictions emerge only when either the science or the beliefs, or often both, encroach into one another's subject matter.

Scientific explanations are based on evidence drawn from examining the natural world, and they rely exclusively on natural processes to account for natural phenomena. Scientific explanations are subject to empirical tests by means of observation and experimentation and are subject to the possibility of modification and rejection. Religious faith, in contrast, does not depend on empirical tests and is not subject to the possibility of rejection based on empirical evidence. The significance and purpose of the world and human life, as well as issues concerning moral and religious values, are of great importance to many people, perhaps a majority of humans, but these are matters that transcend science.

To some people of faith, geology, astronomy, and the theory of evolution are incompatible with their religious beliefs because scientific knowledge is inconsistent with the creation narrative in the book of Genesis and other biblical texts. A literal interpretation of Genesis is indeed incompatible with the gradual evolution of humans and other organisms by natural processes. But that incompatibility emerges only when religious tenets transgress their proper domain. Most biblical scholars and theologians do not consider the Bible to be an elementary textbook of geology, astronomy, or biology; rather, they seek in the Bible religious truths about the meaning and purpose of life and about moral and other spiritual values.

Charles Darwin (1809–1832) is deservedly credited for the theory of evolution. In *The Origin of Species,* his most famous book, Darwin laid out the evidence demonstrating the evolution of organisms. Darwin, however, accomplished much more, and something much more important, than demonstrating evolution.

Namely, Darwin provided a scientific account of the design of organisms, which he accomplished with the discovery of natural selection. The diversity and complexity of organisms, as well as their marvelous contrivances (eyes for seeing, wings for flying, gills for breathing in water) could now be explained as the result of natural processes.

Traditional Christianity had explained the design of organisms as the intentional result of the Creator. Theologians and religious authors argued, for example, that the human eye is as complex a contrivance as a watch or a telescope, with several parts all required to fit together precisely in order to achieve vision. There was, however, a seemingly insurmountable difficulty. If God is the designer of life, whence the lion's cruelty, the snake's poison, and the parasites that secure their existence only by destroying their hosts?

The world abounds in physical catastrophes, such as floods, droughts, volcanic eruptions, earthquakes, and tsunamis that kill thousands and thousands of innocent people. If God had designed the world, it would seem that He would be accountable for these destructive phenomena. Modern science came to the rescue. The scientific revolution of the sixteenth and seventeenth centuries demonstrated that physical events are built into the structure of the universe. The processes by which galaxies and stars came into existence, the planets are formed, the continents move, the weather and the change of seasons happen, and floods and earthquakes occur are natural processes, not events specifically designed by God for punishing or rewarding humans. The extreme violence of supernova explosions and the chaotic frenzy at galactic centers are outcomes of the laws of physics, not the design of a fearsome deity. A person of faith could accept that the world was created by God without the need to attribute to

God's direct action the awesome catastrophes that occur in the natural world.

The scientific revolution of the sixteenth and seventeenth centuries, often called the Copernican Revolution, had left living organisms out of explanation by natural processes. Seemingly, as pointed out by religious authors in the past, organisms give evidence of design, and wherever there is design, there is a designer. Darwin's theory of evolution by natural selection extends the explanation of natural phenomena by natural processes to the design of organisms. In *The Origin of Species,* published in 1859, Darwin laid out the evidence demonstrating the evolution of organisms. Much more important for intellectual history is that *The Origin of Species* is, first and foremost, a sustained effort to solve the problem of how to account scientifically for the adaptations or "design" of organisms. Darwin sought to explain the design of organisms, their complexity, diversity, and marvelous contrivances, as results of natural processes. Darwin brought about the evidence for evolution, because evolution was a necessary consequence of his theory of design. *The Origin of Species* is most important because it completed the Copernican Revolution, initiated three centuries earlier, and thereby radically changed our conception of the universe and the place of mankind in it.

The advances of physical science encompassed by the Copernican Revolution had driven mankind's conception of the universe to a sort of intellectual schizophrenia, which persisted well into the mid-nineteenth century. Scientific explanations, derived from natural laws, dominated the world of nonliving matter, on the earth as well as in the heavens. However, supernatural explanations, depending on the unfathomable deeds of the Creator, were accepted in order to account for the origin and configuration of living creatures—the most diversified, complex, and interesting realities of the world. It was Darwin's genius that resolved this intellectual inconsistency. Darwin completed the Copernican Revolution by bringing the design of organisms into the realm of science, as an outcome of natural processes governed by natural laws.

The Three Failures of Creationism: Logic, Rhetoric, and Science is a pertinent introduction to the logical, philosophical, methodological, and empirical issues that arise in the study of evolution, particularly relevant to readers who may be concerned about the scientific standing of the theory of evolution and how it may relate to religious faith. There is much to learn in this short book, all explained in clear and incisive language.

The book speaks for itself. I will, however, add one further consideration addressed to people of faith. It is my view that attributing the properties and characteristics of organisms to specific design by God is not compatible with faith in the benevolent, omniscient, and omnipotent God of Judaism, Christianity, and Islam. The God of revelation and faith is a God of love and mercy, and of wisdom. A major burden was removed from the shoulders of believers when convincing evidence was advanced that the design of organisms need not be attributed to the immediate agency of the Creator. If we claim that organisms and their parts have been specifically designed by God, we have to account for the cruelty of predators and parasites and for the incompetent design of the human jaw, the narrowness of the birth canal, and our poorly designed backbone, less than fittingly suited for walking upright. Most disturbing is the following consideration. About 20 percent of all recognized human pregnancies end in spontaneous miscarriage during the first two months of pregnancy. This misfortune amounts at present to more than 20 million spontaneous abortions worldwide every

year. Do we want to blame God for the deficiencies in the pregnancy process? Most of us might rather attribute this monumen-
tal mishap to the clumsy ways of the evolutionary process than
to the incompetence of an intelligent designer.

Creationists and proponents of "Intelligent Design" are surely
well-meaning persons. But people of faith would do well to
acknowledge Darwin's revolution and accept natural selection
as the process that accounts for the design of organisms, as well
as for the dysfunctions, oddities, cruelties, and sadism that per-
vade the world of life. As the distinguished theologian Aubrey
Moore already stated in 1891: "Darwinism appeared, and under
the guise of a foe, did the work of a friend. It has conferred upon
philosophy and religion an inestimable benefit." Darwin's theory of evolution is one of the great scientific developments of all
times. People of faith may also see it as a great gift to religion.

Francisco J. Ayala
University of California, Irvine

CHAPTER ONE

Logic, Logical Fallacies, and Rhetoric

In writing this book, it was my intention that it be for people who have no irrevocable position on at least some of the differences of opinion between creationists and evolutionists, but who would like a view of those arguments that is relatively fair. That I have not totally accomplished, as I am clearly an evolutionist and believe in the naturalist (materialist) view, whereas creationists do not. And therein lies the difference. I hope to have produced in this book a clear differentiation of the reasons for what evolutionists believe and what creationists believe, written at a level that intelligent high school seniors or college freshmen or sophomores can readily understand without their having taken any biology or theology courses. I am targeting that group because, in my opinion, it is the failure of scientists to present clearly what they do and why that has caused so many problems in our schools and courts. I welcome criticism from all parties, especially where I have done injustice to any view, and if this book survives to a second edition, I will correct those errors.

I have included a short glossary to aid the reader in understanding some of the terms used in this book.

Not all biologists will necessarily agree with 100 percent of what I have to say. Nevertheless, I believe that the vast majority of evolutionists will agree with almost everything evolutionary that I present. Similarly, not all creationists will agree among themselves that my representation of their view is correct or complete, although the degree to which creationists agree among themselves may often be much less so. The point is that too often one side denounces the other for an opinion that has been given by a member of the opposing camp, even though the opinion being denounced has become rare and unrepresentative of current creationist or evolutionary thought, as the case may be.

Generalities are not intended to be 100 percent applicable, but if something is true 99 percent of the time, that something accordingly is important and frequently not refutable by describing a single exception. This is an example of the *straw man fallacy,* which takes an unrepresentative view of one's opponents and attacks that view, even though it is already recognized as unsupportable in its extreme form by those same opponents.

The principal goal is to establish what science is and how biological evolution is a scientific study, no matter what errors may be present at our current level of understanding of evolution. This is true even if Darwinian evolution itself should be proven wrong. In contrast, creationism, intelligent design, and irreducible complexity are not scientific, even if their conclusions (such as that God made the universe in six literal days about six thousand years ago) were shown to be all correct. It is my hope to represent the creationist viewpoints as those of people with different criteria for resolving important questions. Nevertheless, I hope that if people can understand what evolutionists do and how and why, they will understand that creationism is rarely if ever scientific. Biological evolution is almost always scientific, and thus the reader will understand why evolutionists oppose the teaching of this theological view as part of any science course.

I try to present both sides fairly in describing what strict creationists believe. In evaluating those beliefs, however, I shall rigorously apply scientific and theological methods as appropriate. For example, a creationist may say he believes that the Bible is the word of God and therefore cannot be wrong, that the Bible says the world was created in six days, and that he therefore believes that the world was created in six days. His logical argument in itself is quite valid (and we will elaborate further on chains of logic later in this chapter). Consider the following syllogisms. (A *syllogism* is a form of deductive reasoning consisting of a major premise, a minor premise, and a conclusion.)

Syllogism 1

Premise 1:   The Bible (Old and New Testaments) is the Word of God.

Premise 2:   The Word of God cannot be wrong.

Conclusion 1:   The Bible cannot be wrong.

Syllogism 2

Conclusion 1:   The Bible cannot be wrong.

Premise 3:   The Bible says the world was created in six days.

Conclusion 2:   The world was created in six days.

Note that Conclusion 1 is also the first premise of the second syllogism. But both of these premises are *theo*-logical, not *materio*-logical. They are *theo*-logical because the premises themselves are about God and God's Bible. This difference is crucial in that the starting premises of the creationist are not the same as those of the evolutionist. Hence differing conclusions should be expected, even though our rules of logic are (or certainly should be) identical. The problem of logic is sufficiently important that the majority of this chapter is devoted to logic for those who might enjoy a minimal refresher course on the subject of how we decide which conclusions are logically admissible and which are not.

Critics of evolution often claim that it cannot be correct because it occurs via random genetic mutations, and random processes cannot create order. It is true that mutations are random, because they are not directed by a force that guarantees that, given the ancestral form, the nature of the character can be predicted. But mutations are only half the story. Environmental pressures *are* directive, and this leads to what is termed *natural selection*. Biological evolution is the study of the origins of the diverse nature of living things. It was inspired greatly by Charles Darwin, who, in 1859, proposed a theory for the origin and diversity of the living world. It was, and still is, called *natural selection,* and it postulated that sources of variation (later to be called *mutations*) occurred in nature. Some variants/mutants were harmful and were weeded out. A much smaller number of mutants were beneficial and spread through the population. The mechanism of evolution was that more offspring were produced than the habitat could support, and thus many members of a population would not produce successful, reproductively viable offspring. Those that did, did so because of the favorable effects of the useful mutations. Note that the creation of the mutations is a random process, but *selection of mutants with a beneficial effect is directional.* It was natural to portray the history of life as a genealogy in a branching tree that showed who came from whom and approximately when.

Many devout Christians (and other religious people) find no conflict at all between natural selection and their religion. Evolution, they assert, is simply "the way that God did it." Nevertheless, many evolutionists' statements directly contradict a *literal interpretation* of some of the statements given in the initial chapters of the book of Genesis. For those who can interpret Genesis in a somewhat moralistic, figurative, poetic, or metaphorical rather than a literal fashion, there is no problem. But many people, including "young-Earth" creationists, cannot accept this. Thus, although most creationists are Christian, most Christians are not creationist in the narrow, literalist sense used here. But they could well be creationist in a broader sense—believing, for example, that God started the universe but may subsequently have left it alone to evolve according to His rules. I necessarily differ only from the strict creationists—the literalists.

The controversy between creationists and evolutionists often involves logical failures. For that reason, we begin with a section on logical reasoning and its limitations. Logical failures will continue to be noted.

Logic is the study of the meaning of words and the inferences that are and are not allowable, given some data or reasoning. Rhetoric, on the other hand, although it considers the meaning of words and logic, is not so much interested in what conclusions are true as in what the persuasive effect of the words and gestures may be on you—the reader or listener. The object of rhetorical material is to convince you of something, whether true or not. ("Buy my product," "My client is innocent," "My religion is the only true religion," etc.) Logic will be considered first, then rhetoric.

A. SYLLOGISMS

The study of logic is quite ancient (Aristotle, 384–322 B.C.), but today logic is dominated by truth tables. A truth table is a set of rows and columns showing true/false values for logical propositions and their components. The true/false values are usually shown, in Boolean algebra style, as "1" for true and "0" for false. Perhaps, for an introduction, it is easier to learn a little about syllogisms. A syllogism is an argument. Any one argument is a collection of three statements (sometimes more) of which the first two statements are called the premises (assumptions or givens; we restrict ourselves to three-statement arguments) and the third statement is the "therefore," or "conclusion." The process is of the following form:

Premise 1:	If Socrates was a man, and,
Premise 2:	If all men are mortal,
Conclusion:	Then Socrates was mortal.

The first two statements are the *assumptions* being made, and, *if they are true,* the conclusion in the third line, correctly formed, must also be valid and true. The three lines can be considered more generally as:

Premise 1:	A (Socrates) then B (man).	*OR*	If A is true, then B is true.
Premise 2:	If B (man) then C (mortal).		If B is true, then C is true.
Conclusion:	If A (Socrates) then C (mortal).		If A is true, then C is true.

You will be challenged later to transform an argument into syllogistic form to see if the argument is valid. It is not as easy as one might think. To aid you in achieving success in that effort, some valuable bits of knowledge are presented.

Note that every ordinary syllogism has exactly three different terms or statements: *A, B,* and *C.* Each term is used exactly twice. Term *A* is the subject of premise 1, and *B* is its predicate or result. *B* in turn is the subject of premise 2, with *C* as the predicate. The conclusion eliminates term *B,* and jumps directly from subject *A* to predicate *C.* The order of the two premises is not critical, but the flow of meaning is more natural, and the argument is easier to understand, if the term that occurs in both premises (*B* in this example) is the predicate of the first premise and the subject of the second. Thus:

Premise 1:   Socrates was a man.

Premise 2:   All men are mortal.

Conclusion: Therefore, Socrates was mortal.

Or: *A* implies *B; B* implies *C;* therefore, *A* implies *C* (conclusion). There are many forms of errors in the use of syllogisms, and some of the more common errors are examined at the end of this section.

B. DEDUCTION VERSUS INDUCTION

If one sees that whenever an event happens it is always followed by the same second event, one may come to believe that the first event causes the second one. For example, "Every time I push this button, the doorbell rings. Therefore the pushing of the button is the cause of the doorbell ringing." This is inductive logic. One infers a general rule on the basis of a limited (but generally large) number of observations. However, there is nothing about an induction that guarantees that it will be a correct conclusion. For example, using precisely the same logic, one gets "Every morning just before dawn the cock crows and then the sun rises. Therefore, the cock's crowing causes the sun to rise."

Alternatively, one may conclude (deduce) a particular thing from a general rule. For example, "Dogs eat meat, so I deduce that my Fido eats meat, too." The understood but unstated premise is that Fido is a dog.

Premise 1:   Fido is a dog.

Premise 2:   Dogs eat meat.

Conclusion: Fido eats meat.

The conclusion of a valid deduction *must* be true if the premises are true, whereas an induction *may* be correct but is not proven. The conclusion of a valid deduction is usually of narrower meaning but true, whereas an induction is usually of broader meaning but unproven (although perhaps likely to be true).

The two most commonly used logical forms are deduction and induction. Consider now a series of items that are either logical errors or rhetorical devices designed to convince you, the reader or listener, of something. You need not memorize the names of all these logically related processes, but just recognize them when they occur. It is good practice, if what you are reading feels slippery, to try to find out why you feel that way. You will frequently discover malpractice in word usage. Several such abuses are illustrated here. Writers rarely provide you with a syllogism, or even a partial one, so you must figure it all out for yourself.

C. ANALOGICAL REASONING

Analogical reasoning is the process of making your logic, in a difficult case, exactly like your logic in another case that the listener will readily understand. The point is generally to make the understanding of your logic easier. For example, "If it is not immoral for tigers to eat humans, then it is not immoral for humans to eat humans." The analogical form changes only one word here and invites you, should you believe the first proposition, to accept the second statement as logically equivalent. Most of us would disavow the conclusion, illustrating that analogy does not lead to something having been proved. Its value lies in making it easier to understand the meaning or content of an argument. Darwin said in *The Origin of Species,* "Analogy may be a deceitful guide" (1859, pp. 454–55). Nevertheless, Darwin's first chapter is a long list of analogies arguing that if natural variation is available for the breeder of organisms, variation must be present for natural selection to act upon as well.

Analogies can be humorous, as in, "If practice makes perfect, then *mal*-practice makes *mal*-perfect." This clearly demonstrates that *analogies may make an argument clearer but cannot provide support for an argument.*

D. LOGICAL FALLACIES

1. *Begging the question* (circularity): assuming the conclusion you wish to reach.

A circular argument does not advance our knowledge beyond what was already known or assumed in the premises. That is, the argument being presented begs us to ask the question, *What is the support for the premises?* Or, *What do the premises have to do with the conclusion?* Consider the following syllogism:

Premise 1:   Complex things can be produced only by a designer.

Premise 2:   The human eye is a complex thing.

Conclusion: The eye must have been designed.

Are we sure of the first premise? Is it really true that the *only* way a complex thing can come into being is by the work of a conscious designer?

Or consider how begging the question may be used in political speech:

Premise 1:   Smith is a good family man.

Premise 2:   Smith was a great football player.

Conclusion: Smith will make a good mayor.

We must ask what the premises have to do with the conclusion. Unless you feel that being a good family man or a great football player somehow builds your character or prepares you for political office, the premises have nothing to do with the conclusion.

2. The *equivocation fallacy* (also called a *category error*): using a word with two different meanings in the same argument.

Examples of the equivocation fallacy are given below. A particularly obvious example is the following. First read down the left half of the syllogism to get the silly conclusion that is clearly wrong.

	Silly	*Correct*
Premise 1:	I am a nobody.	I am a person of no importance.
Premise 2:	Nobody is perfect.	There is no individual that is perfect.
Conclusion:	I am perfect.	(No logical conclusion possible.)

Now read the syllogism again utilizing the phrasings on the right for the word *nobody*. This shows how changing the meaning of the term *nobody* between premises 1 and 2 renders the conclusion silly, although the syllogism on the left side is *formally* valid.

The usage of this equivocation fallacy is the basis of the creationist's rather insidious declaration that evolution is not a fact. One meaning of the word *theory*, found in ordinary, everyday usage, is that of a guess. Creationists often say in a disparaging tone that evolution is only a theory, meaning that it is only a guess and not a fact. But when a scientist uses the term *theory*, the scientist means a well-supported explanation uniquely consistent with many thousands of observations. Consider, for example, Newton's theory of motion, Copernicus's heliocentric theory (that the Earth rotates around the sun), Einstein's theory of relativity, or atomic theory.

The creationist's syllogism goes like this:

Premise 1:	Evolution is a theory.
Premise 2:	A theory is only a guess.
Conclusion:	Therefore the theory of evolution is only a guess (and thus is not a fact).

This argument is invalid because it uses *different meanings of the word* theory. Its illogical result arises from changing the meaning of the word *theory* between premise 1 ("a well-substantiated scientific truth") and premise 2 ("a guess"). This has also been called the four-term fallacy because of the two different meanings of *theory*. The correct syllogism is:

Premise 1:   Evolution is a theory.

Premise 2:   A theory is a well-supported explanation of many observations.

Conclusion: Therefore, evolution is a well-supported explanation of many observations.

This form of the argument is valid and the premises are true, so the conclusion is true. It is puzzling that creationists continue to assert that evolution is "only a guess."

A third instructive example of the equivocation fallacy is the following:

Premise 1:   Humans are a species.

Premise 2:   A species is a mental construct.

Conclusion: Humans are a mental construct.

The *concept* of a species is a mental construct, a creation of the human mind, but that is different from human beings being a species. To make a valid argument, the statements should be as follows:

Premise 1:   Humans are interfertile with other humans. (That is, they can interbreed.)

Premise 2:   A group of interfertile individuals is termed a species.

Conclusion: Humans are a group called a species.

3. The *excluded-middle fallacy:* assuming that only two alternative explanations exist, black or white. (E.g.: if Darwinism is false, then Genesis must be true.) There is no allowance for a gray or intermediate theory. This is a very important fallacy in the discussion between scientists and creationists. Creationists often construct a world with only two arguable views: creationism, and materialism/naturalism. There is no recognition that more than one version of creationism exists, such as "young-Earth" versus "old-Earth" creationists. Similarly, there is no recognition that more than one version of naturalism exists, such as Lamarckism and Darwinian evolution. This produces a situation in which there are, seemingly, only two contending theories, one of which then must be false and the other must be true. Such circumstances are wonderful but rarely occur in nature, because by proving that one of the contenders is wrong (the evolutionist position), one appears to prove that the other contender is necessarily right (the creationist position). Unfortunately, this argument has a major logical flaw. That flaw is that one is not allowed, arbitrarily, to omit some explanations, or to tie several of them together to get a single entity, in order to reduce the logical system down to a maximum of only two components.

For example, how can the synthesizing machinery obtain the correct sequence of amino acids in a protein? The creationists have their choice: model 1 (God does it); and model 2, a random model for which they can calculate probabilities. Model 2 is the creationist view of what *they think* evolutionists believe. The creationists make the calculations and rightly reject these random models. Having rejected model 2, they then infer that the creationist model, model 1, must be right. But the inference requires that there be no third model, random or not. In fact there are an infinite number of possible different models. The conclusion depends upon there being no other random models, including a model 3, which is discussed later on (see chapter 3, section B). More than two possible models are conceivable. Thus the creationist argument is already a failure because it is limited to only two possible models. But it gets worse. The evolutionist argument for model 3 has a very high probability of forming the whole protein without error.

4. The *genetic fallacy:* arguing against an idea on the basis of the proponent's personal character. An example is: "That man is a natural born idiot. How could any self-respecting human vote for his proposal?" [See ad hominem]

5. The *naturalistic fallacy:* asserting what ought to be true on the basis of what appears to be true. A disquieting example of this fallacy: Some animals eat their young; therefore it would be OK if humans were to eat theirs. This can be stated as a syllogism as follows:

Premise 1: It is acceptable for humans to do what some animals do.

Premise 2: Some animals eat their young.

Conclusion: Therefore, it is acceptable for humans to eat their young.

Premise 1 is disquieting but not necessarily in itself false. Premise 2 is true. The conclusion appears to be true, so the argument is valid. But the argument is not morally acceptable. In the early nineteenth century, the social Darwinists (see the discussion in chapter 2, section B) accepted all this and tried to sterilize mentally disabled people and deny them schooling and welfare etc. on the grounds that the intended recipients were inferior, by nature, and so could not benefit from education. Logic fails us here, but only because ethics and morality are not considered. Logical deductive reasoning can tell us when certain facts are true in terms of a "truth table," but it does not prescribe how those facts should be used. The conclusions drawn from a logical, abstract evaluation may be unacceptable in moral terms or irrelevant to our daily lives (e.g., an imaginary situation). The term *naturalistic fallacy* was originally defined (by G. E. Moore) somewhat differently from how the current popular definition describes it. The naturalistic fallacy is summarized more frequently by saying that we must not assume an "ought" from an "is." As Albert Einstein wrote, "For the scientific method can teach us nothing beyond how facts are related to, and conditioned by each other.... knowledge of what is does not open the door directly to what should be" (1954, pp. 41–49).

The Italian cardinal Caesar Baronius (quoted by Galileo in his letters) summed up the separation of ethical philosophy from scientific study in 1598 with a wise remark; "The Bible was written to tell us how to go to heaven, not how the heavens go." In the Bible (Luke 20:24–25), a commission of hostile priests challenges Jesus on the issue of authority, asking him whether it is lawful for a believer to pay taxes to a secular authority such as Caesar. Jesus's response is:

> "Shew me a penny. Whose image and superscription hath it?" They answered and said, "Caesar's." And he said unto them, "Render therefore unto Caesar the things which are Caesar's, and unto God the things which be God's."

We could do worse than to consider ethics and morality to be God's domain, and natural science to be Caesar's.

6. The *non sequitur* (Latin for "does not follow") *fallacy:* asserting that a conclusion follows from the preceding material when it in fact does not. For example: "That critical reviewers advance scientific arguments against intelligent design (whether successfully or not) shows that intelligent design is indeed falsifiable" (Behe 2000). To see the incorrectness of the argument, replace the words *intelligent design* in the syllogism below with the word *God,* which leads to the conclusion that God is falsifiable. All fallacies are, in one way or another, non sequiturs.

The argument, expressed as a syllogism, is as follows:

Premise 1:   Intelligent design has been argued against by scientists.

Premise 2:   If scientists can argue against (or for) intelligent design, it is falsifiable.

Conclusion: Therefore, intelligent design is falsifiable.

To be considered scientific, a theory must be falsifiable—that is, you should be able to prove the theory false if a particular fact is observed; otherwise, you have no means of testing the theory by experiment. In the syllogism above, the argument is valid in that the conclusion would be true if the premises were true (Tymoczko and Henle 2000). It is not clear whether the conclusion is true or not. If the conclusion was true, intelligent design could be regarded as scientific, thereby destroying a barrier between evolution on the one hand and creationist areas of knowledge, science, and theology on the other.

7. The *rationalistic fallacy:* believing that rational arguments will persuade—or, the assumption that human beings will govern their affairs on a purely rational basis by using only logical trains

of reasoning. This is a common error, especially among professors; this book may be an excellent example of it! Rational arguments may not persuade if they are difficult to follow or if they challenge long-held and cherished beliefs.

8. *Reductio ad absurdum* (Latin for "reduced to an absurdity"): reducing an argument to the point of making it appear absurd. *Reductio ad absurdum* makes use of the law of non-contradiction, which says that a particular statement "A" cannot be both false and true at the same time. Of course, the demonstration of absurdity may rely on a very extreme example of the principle being discussed—so extreme that it is an unfair distortion of the statement given. (See fallacy 9, "Straw man," below.) The wording of the statement is very important, particularly when you use statements like "*A* is *always* true" or "*B* can *never* occur." A single counterexample can show your position to be absurd. Mark Twain made use of the reductio ad absurdum principle when he wrote that many years ago the Mississippi River was "upwards of one million three hundred thousand miles long." (See chapter 2, section E: "Mark Twain and Science.") Twain is illustrating the absurdity of assuming that the rate at which the Mississippi River is seen to shorten today has remained constant over millions of years.

9. *Straw man:* representing an opponent's view in a form so extreme that it is absurd. It suggests that your opponent's logic must be bad when the only thing that is proven is that the argument does not hold in the extreme. For example, your opponent gets his answer by dividing by a very small number. You then declare him wrong because, if you divide by zero, which is very small indeed, the answer is undefined. (See fallacy 8, "Reductio ad

absurdum," above.) Demonstrating a contradiction in an argument is valid, but not if you distort the argument into a case it was never intended to cover.

10. The *tautological fallacy:* formulating a conclusion that is true for whatever set of values (*true* or *false*) are entered into our table. The terms are defined in such a way that the conclusion cannot be disproved. Examples of tautologies are "the law of the excluded middle" (*A* or not-*A*), "de Morgan's law" (if not both *A* and *B*, then either not-*A* or not-*B*), and "proof by cases" (if at least one of *A* or *B* is true, and each implies *C*, then *C* is also true). Tautologies are not always fallacious. The tautological fallacy occurs when the conclusion is already contained in the premises—perhaps using slightly different words. The logical argument does not advance us beyond what is already known or assumed.

11. *Miscellaneous fallacies.* Either or both premise lines may not be true, and even the conclusion line may not be true. If the first two premise lines are true, then the conclusion line should be true and will be as long as the argument is valid. For example, recall the syllogism at the beginning of section B, "Deduction versus Induction," above:

Premise 1:   Fido is a dog.
Premise 2:   Dogs eat meat.
Conclusion: Fido eats meat.

This syllogism is sound. But what of the case where the third line instead reads "Bears eat meat"?

Premise 1:	Fido is a dog.
Premise 2:	Dogs eat meat.
Conclusion:	Bears eat meat.

The argument is not valid even though the conclusion is true. The third line introduces a forbidden fourth term ("bears"), and hence this case is also called the *four-term fallacy*. This represents an interesting case in that both premises and the conclusion ("Bears eat meat") are true *even though the logic—the argument—is not valid*. Although the formula is true in terms of logic, it was just a coincidence that bears actually do eat meat; it didn't necessarily follow from the premises. To see that the conclusion does not actually flow from the premises, consider another syllogism with the four-term fallacy:

Premise 1:	Polly is a bird.
Premise 2:	Birds have feathers.
Conclusion:	Bears have feathers.

The logical structure is similar, but in this case you can see that the conclusion is absurd.

E. RHETORICAL DEVICES

Rhetorical devices use various phrases and tones for their effect, with or without regard to logic. "What is the authority for . . . ?" is a rhetorical question that implies that, after considerable search, no such authority will be found. All of the fallacies, if they are used despite the user's knowing they are fallacies, are then rhetorical devices. Note the famous phrase "That's a rhetorical question," meaning that you aren't supposed to answer the question, which was posed only for effect. Loaded words are another, and very common, rhetorical device.

1. *Ad hominem:* attacking the speaker rather than the speaker's argument. It is rather in the spirit of "If you have no good arguments on the basis of the facts, then you should: (1) cause confusion; (2) shout louder; (3) assert your opponent's ignorance of the issue; (4) accuse her of unethical or immoral acts; (5) ridicule your opponent; and so forth. Some recent examples include the following:

a. The astronomer Fred Hoyle has hypothesized that life may not originally have begun on Earth, but began somewhere else and then migrated to our planet (by various interesting ways). Daniel Dennett observes, in his book *Darwin's Dangerous Idea* (1995, pp. 314, 318), that skeptics sometimes refer to this idea as "Hoyle's Howler" as a way of insulting Hoyle by implying that he is stupid.

b. "The fact that a distinguished philosopher overlooks simple logical problems that are easily seen by chemists suggest that a sabbatical visit to a biochemistry laboratory might be in order" (Behe 1996, p. 221; Behe is a creationist). This sarcastic remark is also insulting.

c. "I have encountered this blunder so often in public debates that I have given it a nickname: 'Berra's Blunder'" (Johnson 1997, p. 63). This is in the same category as the howler. Phillip Johnson was referring here to Tim Berra's use of the changing automobile design in the Corvette sports car to illustrate the concept of "descent with modification." (See chapter 2, section G.1, "Automobile evolution.")

d. "...creationist canards (lies) [regarding thermodynamics]..." and "...these thermodynamics howlers..." written by Paul R. Gross (an evolutionist) in his review of a book edited by Matt Young and Taner Edis entitled *Why Intelligent Design Fails: A Scientific Critique of the New Creationism.*

Note that these examples of attacking the person rather than the scientific claims as exemplified are used by both creationists and evolutionists (two each of the four examples). It is reprehensible whichever side does it.

2. *Ad ignorantiam:* using the ignorance of one's opponent as evidence of the correctness of one's own position. (See rhetorical device 1, "ad hominem," above.)

3. *Loaded words.* "We all have naturalism in our bones and even conversion [to Catholicism] does not at once work *the infection* out of our system." (Citing of C. S. Lewis by Dembski; emphasis mine.) *Infection* is an excellent example of a loaded word, as is easily demonstrated by replacing *infection* with a neutral word. For example: "We all have naturalism in our bones and even conversion [to Catholicism] does not at once work *our prior beliefs* out of our system." (The difference between the two statements is emphasized in the italicized words.)

"Methodological atheism": Phillip Johnson is talking about the scientific method, which scientists use and which may reasonably be said to be naturalistic. Creationists have no comparable method, which frequently hurts the creationist arguments where the issue is one of whether intelligent design is or is not scientific. Thus, if you insinuate that the scientific method is atheistic, you tend to reduce the importance of the scientific method in the reader's mind. Since naturalism (or materialism) is logically independent of theology (see chapter 2, section B.4, "Logic/epistemology"), their mixing is particularly loose.

Some creationist advocates have favored the term *creation science* as a means of suggesting that it, too, is scientific. Evolutionary scientists in return have scornfully referred to creation science as an *oxymoron*—a loaded term if there ever was one. (An oxymoron is a term that is inherently self-contradictory. Notable examples include "deafening silence," "civil war," "friendly fire," "jumbo shrimp," "original copy," and, yes, some people maintain, "military intelligence." Students at Occidental College in Los Angeles have any number of "Oxymoron" jokes.) Certainly the level of emotion in our example would be greatly reduced by saying instead that creation science is not in fact scientific. (See "Rhetorical Devices," in this section)

Another example of loaded words is the following humorous conjugation of verb forms such as "I am persevering, you are stubborn, he is pigheaded."

Examples of loaded words can be seen in a discussion of peppered-moth selection involving a creationist (Jonathan Wells) and two evolutionists (Kevin Padian and Alan Gishlick). Wells wrote a book, *Icons of Evolution: Science or Myth?*, that Padian and Gishlick reviewed. The data at issue are from peppered-moth studies carried out by H.B.D. Kettlewell.

Kettlewell's research is about moths that are generally peppered or very light in color. They spend most of their lives perching on the bark of trees. In the mid 1800s, when the industrial revolution was occurring, industry smokestacks were emitting much soot and thereby blackening the trees in industrial British cities like London and Manchester and their neighbors. And about that time someone found a previously unseen dark moth. As the trees got blacker, the frequency of dark moths increased, reaching sometimes to 98 percent. With an interest in preserving the environment, laws were passed to reduce the pollution. To no surprise for a Darwinist, the frequency of the dark moths declined again as the blackness of the trees declined.

How might this have occurred? Birds are known predators of these moths, and it was soon suggested that the increase of dark moths was a matter of camouflage. When the bark of trees was black, dark moths were difficult for the birds to see, but when the bark of trees was whitish, peppered moths were the variant that were difficult to see and thus their chances of survival were enhanced. The positive correlation between the blackness of the trees and the frequency of dark moths supports the proposition of natural selection going on before your eyes. Kettlewell illustrated the camouflage by pinning a peppered and a dark moth side by side on a dark tree trunk and also a similar pair on a light-colored tree. It was astonishing how well the dark moths blended in with the blackened trunk, and equally astonishing how the peppered moths blended in with the light-colored trunk.

But things are not quite as simple as they may at first appear. The Kettlewell study was incomplete in that it failed to properly consider other possible factors, such as migration of moths from surrounding areas that could have overwhelmed the influence of selection. The Kettlewell study also gave undue emphasis to moths resting on tree trunks and failed to consider that birds see ultraviolet light much better than humans and thus might have been able to detect moths that are well camouflaged to human eyes. There are unanswered questions, but the evidence for differential survival in agreement with the selection hypothesis is basically sound, despite the incompleteness of the Kettlewell study.

Wells published a rejoinder to criticisms raised by Padian and Gishlick—criticisms that included pejorative phrases like "*notorious* peppered moth experiments," "*staged* photos of moths on tree trunks," and "the statistic is *bogus*." (Emphases mine.) The statistic arises from the observation of forty-seven peppered moths observed resting in the wild, of which twelve were resting on a tree trunk, giving $12/47 = 0.225$ of the resting peppered moths located on tree trunks. This was hardly the critical measurement of the study, but it can be said to demonstrate the assertion that peppered moths do rest, in sizable numbers, on tree trunks. That is important, not bogus. In conclusion, loaded words should not be used to attempt to sway your audience.

4. *Repetition.* "Testing Darwinism by the molecular evidence has never been attempted.... The true scientific question—Does the molecular evidence as a whole tend to confirm Darwinism when evaluated without Darwinist bias?—has never been asked." Repetition is a form of emphasis present in the two phrases "has never been attempted" and "has never been asked." Moreover, "Darwinist bias" would be unloaded were it altered to "the Darwinist view."

F. OTHER TERMS RELEVANT IN LOGICAL ANALYSIS

1. *Bias:* Any assumption, often unrecognized, that tends to cause the experiment to produce inaccurate answers, pushing the results in one direction. (See *objectivity* and *subjectivity* below.)

2. *Objectivity:* A scientist's goal, reflecting the scientist's attempt to see what is there in his experiments rather than what he hopes,

believes, or expects is there. It is the overcoming of one's personal biases or inclinations. This is something that is often difficult to achieve. A common phrase is "If I hadn't seen it, I wouldn't have believed it." This typifies the nature of objectivity. A different humorous phrase, also typifying objectivity, is "If I hadn't believed it I wouldn't have seen it." Shakespeare seems to have recognized the problem:

> HAMLET: Do you see yonder cloud that's almost in shape of a camel?
>
> POLONIUS: By the mass, and 'tis like a camel, indeed.
>
> HAMLET: Methinks it is like a weasel.
>
> POLONIUS: It is backed like a weasel.
>
> HAMLET: Or like a whale?
>
> POLONIUS: Very like a whale.

Wishful thinking can often lead us to accept "evidence" that would be rejected by a more objective observer. For example, a primitive human called "Nebraska Man" was once thought to have existed, based on the evidence of a tooth. It was found later that the tooth was not from a human but from an extinct peccary (a piglike hoofed mammal), and had been misidentified as being primitive human. Other examples include the "Paluxy Event" and the "Piltdown Affair." (See the discussion in sections I and J of chapter 4.) In a more famous example, Martin Fleishmann and Stanley Pons claimed to have produced "cold fusion" in 1989—but this claim has not been accepted by the scientific community. A case of wishful thinking? In science, it is best to proceed with a good dose of humility.

3. *Subjectivity:* The misreading of evidence because of personal beliefs. (See *bias* and *objectivity* above). Problems associated with

bias, (lack of) objectivity, and subjectivity are common to evolutionists and creationists alike. But the two groups typically do not respond in the same manner.

Suppose there is disagreement over two proposed dates for the age of some event or artifact. The evolutionist, understanding the requirements of the scientific method, will ask whether his determination of the dating is repeatable on another sample from the same geographic site. The creationist, on the other hand, doesn't have much interest in repeatability except for its hoped-for conclusions. Then the lack of scientific repeatability is, for the creationist, evidence that the science is wrong and creationism is right. Of course, in the worst case for the scientist, both dates are quite wrong, whereas in the best case for the scientist, both dates are correct but with larger-than-hoped-for error bars. An extreme example of this sort is the date for the Earth's origin.

4. *Relevance:* the appropriateness of an argument for the question being asked.

5. *Moot:* no longer relevant. For example, the question "Did any dinosaurs survive their great extinction at the end of the Cretaceous 65 million years ago?" would become moot if a fossil dinosaur were to be found that dates to more recently than 65 M.Y.A.

6. *Implication:* a proper conclusion, given acceptance of the prior assertions (the premises). Whenever it is impossible for *A* to be true without *B* also being true, it is said that *A strictly implies B.* Although the word *entails* is sometimes used as a synonym for *implies,* some logicians (notably Alan Ross Anderson and Nuel D. Belnap) have argued that for *A* to *entail B,* not only must it be

impossible for *A* to be true without *B* being true, but there must be some *relevance* between the truth of *A* and the truth of *B*. For example, a contradiction *implies* the truth of *any* proposition whatsoever: "Wolves eat meat and wolves do not eat meat" logically implies "The Earth was created six thousand years ago"; yet it entails only those propositions that are relevant (e.g., "Wolves eat sheep").

7. *Invalid:* not having the proper structure of a syllogistic argument. When the conclusion does not follow from the premises, the syllogism is said to be invalid. Even if the conclusion is true, and even if it is an observable fact, if the structure is not proper, the logic is invalid. It is possible to have a true conclusion in an invalid syllogism. The conclusion may be true, but since it does not follow from the premises, the syllogism is invalid.

CHAPTER TWO

The Basics

What is "basic" to the understanding of creationism and evolution? This chapter discusses some of the categories of knowledge and belief, and examines areas of knowledge and information.

A. HOW DO I KNOW ANYTHING?

I suggest seven ways of knowing, not all of which are equally dependable. Examining these ways can be critical in deciding the logic or correctness of a conclusion.

1. *Experience* is intended to cover the effects of a lifetime of living in the world and learning that if we're hungry and cry, mother will give us milk; if we stand in the rain, we're likely to get wet. Every time we throw our toy out of the crib, it falls on the floor and we begin to learn about gravity and balance. Through the use of induction (reasoning from specific facts to general rules), we begin to formulate laws that seem to us to govern events in

our world. Our experiences are formed through our perceptual senses: seeing, hearing, touching, smelling, and tasting. Unfortunately, our perceptions are fallible. The philosopher René Descartes argued that our observations may be due to a dream, a deceiving God, or a deceiving daemon. Philosophers belonging to the branch of philosophy known as *skepticism* point out that we cannot be sure that we are not a disembodied "brain in a vat" and that all our observations are merely illusions being fed to us. (Think of the movie *The Matrix.*). Various justifications have been proposed for thinking that our experiences are reliable. One answer to the problem is the formation of a suitable track record of memories that makes it reasonable to attribute reliability to our perceptions and gives us some hope that our observations correspond to reality!

2. *Observation* is like experience but more structured. You perform experiments with proper controls. You time how long it takes to get to work by different routes and then use the one that gets you to work the quickest—unless, of course, you dislike work. You alter the spices in a recipe to see if the resulting pancakes taste better or worse. Scientists continually use this method to learn about the material world, and for them it is the definitive way of knowing that some belief (knowledge) is correct. There is no higher authority to which a scientist can appeal. It is the most important source of information for scientists, because it is verifiable by additional and repeated observations. One does need to be careful, though, because things are not always what they superficially seem. Moreover, as someone once said, "If I hadn't believed it, I wouldn't have seen it." This important humorous saying indicates that it is possible to be led astray by believing

something so strongly that you are led to see things that are not there, or to fail to see things that are.

3. *Logic* is the mathematical subject that assures that the reasoning process is valid, with erroneous reasoning revealed as such. But logic doesn't prove anything even if the logic is valid, because the correctness of the conclusion still depends upon the correctness of the assumptions. All knowledge can be put through the tests of logic, logic having nothing to do with the philosophy one may be examining. It can test for logical (in)consistency. Thus both creationism and science (and other systems of belief) strive to make their knowledge system as logical, as internally consistent, as possible. It is the second most important way of knowing for the scientist, but logically it is important for the creationist as well.

4. *Authority* is learning from the learned, the most bountiful source of knowledge that we have. We learn first from our parents, then from teachers and playmates, and then from employers. We also learn from books, newspapers, magazines, television, and games. Not all authorities are equally good, and a good authority in one field may be limited in another field. If you have a theological question, you'll probably do better to ask a rabbi (priest, imam, etc.) than a scientist. But if you have a scientific question, the scientist is probably a better choice than the rabbi (or priest or imam). And even in their own fields scientists are not all equally knowledgeable. As the Romans put it: *caveat emptor*—let the buyer beware.

For the literal creationist, on matters of evolution there is no higher authority to which one can appeal than the first two

chapters of Genesis, literally interpreted. On this the creationists insist. It should come as no surprise, then, that, as things now
stand, the differences between evolutionists and the strict creationists are irresolvable. They could be resolvable only if one
of the two groups changed their criteria for truth to that of the
other, or if they separated the areas addressed between them.
Specifically, the evolutionists would agree that they have no
authority in the realm of the theological, and the creationists
would agree they have no authority in the realm of the material
world. One should recognize that authority comes in different
flavors. It is important that one recognize the difference between
an authority that talks a lot about what is true, and one that
presents experimental evidence in favor of a proposition being
true (or not).

5. *Intuition* is the sudden appearance of an idea that feels correct
to you although you can't say why or how it came into your head.
It is a frequent source of knowledge, although its probability of
being correct may be very low. "I don't know whether this organism (Archaeopteryx) is a bird or a mammal, but I feel that it is a
bird." Scientists frequently have these feelings, and those who
have the best intuition tend to be the more successful scientists.
Intuition tends to be most valuable when it is about a topic of
which the one doing the intuiting has much experience. Note
how weak the prediction is. Intuition is of little value in an intellectual argument, but it may be of great value in suggesting a
fruitful line of research.

6. *Revelation* is God speaking to you, telling you what to believe
or do, or what is true. For scientists it is their experiments or

observations that are revealing, rather than God. The scientist's attitude is often one of "If God tells you something, that is revelation. When you tell me what God said, that is hearsay."

7. *Faith* is the knowing of something for which none of the above applies: "There is no logical reason why I believe this, but I am certain that it is true." The philosopher Søren Kierkegaard was of the opinion that there are always gaps in what we can determine by observation and logic. When so-called expert witnesses are called into the courtroom, they may disagree as to what the facts are. Many experiments have been performed to demonstrate that eyewitnesses can often disagree as to the details of an event that they all witnessed. A famous experiment invited observers to watch a video and count the number of times a basketball is passed back and forth. During the video a person wearing a gorilla suit strolls onto the court, but the vast majority of observers are so intent on counting the number of passes that they fail to notice the gorilla! The shortcomings of observation and logic led Kierkegaard to state that embracing a faith such as Christianity requires a "leap of faith." But even Kierkegaard noted that this leap is taken with "fear and trembling"—that is, we can never be certain that our faith will lead us in the proper direction.

Each of these ways of knowing can sometimes serve a useful purpose, but for the scientist as a scientist, only careful, controlled observations can decide between two contradictory materialist views. The paradigm is to discover what contradictory predictions the two views make and how to discover data and perform experiments that will give results that determine which of the opposing views, if not both, is clearly incorrect. A scientist's explanation must function; that is, it must permit control over some observed condition of the material world. A creationist's explanation for the same observations need only be asserted.

B. FOUR AREAS OF KNOWLEDGE

1. *Theology* (metaphysics), as used here, is the study of gods and their activities, which leads to questions such as "How many gods are there?" and "Are any of them male?" and "How many angels can dance on the head of a pin?"

2. *Ethics* is a system of moral principles to guide human conduct. It deals with an individual's standard of conduct or a body of rules pertaining to social obligations and duties. The word *ethics* comes from the Greek *ēthikos,* meaning "personal disposition." Much has been written concerning which principles a person should use to guide his or her personal conduct. It has been claimed that the theory of evolution destroys faith in the Bible as an authority for moral guidance, and hence is an attack on morality and ethics. Although the theory of evolution differs from literal interpretations of the Bible on factual questions such as the age of the Earth, evolution says nothing about what moral codes we should follow. On this matter, evolution is currently silent or neutral. In addition, moral codes can change over time and are derived from a number of sources other than the Christian Bible. Confucius devised a system of ethical precepts based upon the practice of *jen* (sympathy or human-heartedness) centuries before Christ. In ancient Greece, idealists such as Plato held that there is an absolute good to which human conduct aspires. Ethical systems have been ascribed to divine will, but also to an innate sense (Jean-Jacques Rousseau) and to human experience (John Stuart Mill and John Locke). David

Hume made contributions on the nature and necessity of a humanly inspired morality. Immanuel Kant sought to set up an ethical system independent of theology, and spoke of the categorical imperative: "Act as if the maxim from which you act were to become through your will a universal law." Philosophers such as G.E. Moore have postulated an immediate awareness of the morally good.

Science does not pretend to answer moral questions. But science has wrongly been cited as a justification for moral and ethical views, which is a misuse of the theory of evolution. As Fran de Waal has rather cynically observed: "Any framework we develop to advocate a certain moral outlook is bound to produce its own list of principles, its own prophets, and attract its own devoted followers, so that it will soon look like any old religion."

The theory of evolution has frequently been cited by various people to support their economic and political goals. I will refer to this group, loosely, under the rubric of "social Darwinists." This is a loose and imprecise term, because the goals and positions of those who have been called, or called themselves, social Darwinists have differed greatly. In this work, I refer to "social Darwinists" as those who profess to determine an ethical *ought* from an empirical *is*—that is, those who claim to know which ethical path is best for society from contemplating nature and the theory of evolution. Such social Darwinists profess to "help nature along" by speeding up the time needed for "progress." They have professed to know which economic system is best with the "certainty of the principle of gravitation" (David Ricardo). They have professed to know which people in society ought exclusively to reproduce (Francis Galton, Harry Laughlin, Margaret Sanger) and which races are superior.

Accordingly, they have advocated the sterilization of the unfit, opposed state aid to the poor and state support of education, and advocated permitting unrestrained business and commerce. Such beliefs have used evolution and nature to justify racism, colonialism, slavery, and Nazism. Sir Francis Galton wrote, "What nature does blindly, slowly and ruthlessly, man may do providently, quickly, and kindly. As it lies within his power, so it becomes his duty to work in that direction." This citation of nature's actions as a basis for our own is an excellent example of the naturalistic fallacy, which we covered earlier. It is also an abuse of analogy arguments. Moreover, even if one wants to improve the human condition, nothing in the theory of evolution indicates that it would be helpful to turn the strong against the weak.

There is no evidence that nature or the theory of evolution demands any specific moral code. Evolution requires only that species survive long enough to reproduce. Nature should not be personified as someone who is anxiously wringing her hands, watching events unfold, hoping that the "good" side will win, that the "bad" side will be defeated, and that "progress" will be made. Nature has no grand teleology: it is not working toward any ultimate goal. Nature makes no judgments as to the particular method of survival that species employ. Many strategies are employed by various species, and no particular strategy can be considered paramount in the attempt of species to survive. Many social Darwinists (e.g., Herbert Spencer) emphasize the competitive aspect of nature, but such writers as Peter Kropotkin have noted that many species survive by means of cooperation. Parental self-sacrifice is example of altruistic behavior. Various traits can be observed in nature, but these traits differ widely and can be polar opposites. Each trait is appropriate to the particular species, giving advantages to the species member or group for its particular environment and morphology. But which of nature's many traits is most advantageous for humans to cultivate?

It is pointless to arbitrarily select one trait from nature to be used as a guide for society's path, because it is impossible to know which particular trait will ultimately be the most important in the game of survival—to say nothing of the injustice of imposing such a questionable decision upon others. Developing strength and power might seem to be a good strategy in the game of survival, yet the large and powerful *Tyrannosaurus rex* went extinct—probably due to an inability to adapt to a cooler environment. Studies have shown that when a poison (warfarin) was introduced into the environment, the local rat population adapted so that poison-resistant rats became dominant. When the poison was withdrawn, the resistant rats proved to be not as well adapted to the nonpoisoned environment as the poison-susceptible rats. The poison-susceptible rats quickly became dominant again. The rats that were "fit" for one environment proved to be "unfit" for another environment. So, how is one to judge which trait makes a species member most "fit"?

Of course, we can select an arbitrary trait that has become more pronounced over a period of time and make the claim that the development of this particular feature is what nature is "trying" to accomplish. But, as noted later in this chapter, in the discussion of Mark Twain and the Mississippi River, it is a mistake to assume that a short-term trend will continue or ought to continue indefinitely. For example, the average height of humans has been increasing in recent years. From that data, should we assume that nature "wants" to produce ever-taller men and women or that an average height of fifteen feet would be superior or desirable? There is some evidence that a greater height correlates with economic success. But height can also be disadvantageous. Longer limbs are more easily broken. A tall person who falls is more likely to be injured. In certain sports, such as gymnastics, it is desirable to be short so as to have a low center of gravity. Since height can be affected by environmental conditions as well as genetics, the social Darwinists have their work cut out for them in trying to separate environmental factors from genetic factors, and determining if a certain trait really is an advantage.

In addition, it is physically impossible to pursue every trait simultaneously. It would be like trying to breed an animal as large as an elephant, as fast as a cheetah, and with the flying skills of an eagle. Certainly each trait of each animal has it uses. But the advantage of the elephant's strength, or the cheetah's speed, or the eagle's flying skills depends upon the particular situation. Cockroaches are known to be able to withstand radioactivity to a great extent. If humans get themselves into a worldwide nuclear war and contaminate the planet with radioactive debris, it may be that cockroaches take control of the planet from humans! A strange future from the human point of view, but nature (to anthropomorphize) would be "perfectly content" with that outcome. We cannot rule out the possibility that the attribute of the cockroach to withstand radiation may one day prove to be the trait that is most useful! On that basis, the social Darwinists should advocate that we attempt to be more like cockroaches. To be sure, we can arbitrarily select a particular trait and try to cultivate it by artificial selection. But to disregard the moral issues involved and claim that the cultivation of a particular trait is what nature "wants" us to do with the certainty of the principle of gravitation is absurd.

The social Darwinists really have a hopeless task in their professed ambition to determine where nature is going and to "help nature along" by speeding up the process. As the theologian Alan Watts has noted, those who claim to know the right course of action by the contemplation of nature are in the paradoxical situation of compelling nature to follow its own laws, as if to say, "Dammit [nature] why can't you be more natural!"

3. *Esthetics* is the study of the beautiful and the ugly and the fine arts (music, painting, sculpture, theater performances, etc.), which leads to questions such as "Is a rhinoceros beautiful?" and "Was an exploding atom bomb over Kwajalein [a South Pacific island where the atomic bomb was tested after the island's inhabitants were resettled] beautiful?" "Was it more or less beautiful over Hiroshima?"

4. *Logic/epistemology* is the study of matter and energy and their interactions, which leads to such questions as "Why is the sky blue?" and "Why does a compass needle point north?" and "Why does a straw look bent when stuck in a glass of water?"

We make no claim that there are only four nonintersecting areas of knowledge. (Is "the set of all numbers" an area of its own, or should it be categorized as part of the "materialistic" area, or as some other area?) For our purposes it is sufficient that simple theological and epistemological/materialistic items cannot be members of the same area. Note that these four areas were chosen to be broad but mutually exclusive. Nearly every simple piece of knowledge appears to be assignable to one of the four areas. More complicated knowledge may have simple forms that are combined. Thus religion (as opposed to theology given above) is complex, having all four (at least) areas represented in it. Religions have an interest in the number of Gods there are (theology); an interest in the moral behavior of their members—for example, the Ten Commandments (ethics); an interest in their house of worship (mosque, temple, cathedral, etc.) being a beautiful place in which to praise God and encourage the congregation to come to worship (esthetics); and an interest in the epistemological study of material things, needing to understand physical laws so that their house of worship will not collapse in a big storm.

The point of this little exercise is that, at the simplest level, each subject is completely independent of the other three. Thus, for example, the one Christian God seen in Michelangelo's painting on the ceiling of the Sistine Chapel is beautiful (esthetics), but that beauty cannot tell us whether there is only one God (theology). More particularly for us, it indicates that science, which is confined to the logical/epistemological study of the material world, cannot answer questions of morality, beauty, or theology. This does not prevent the occasional scientist from trying to use the deservedly great reliability of science to claim that God does not exist. On such a matter, science can only be agnostic, honest, and note that science has no authority in the theological realm. But similarly, theology has no authority in the area of the epistemological study of the material world. There is some irony in the painting on the ceiling of the Sistine Chapel of God giving life into Adam by the touching of fingertips when Genesis 6:7 says, instead, "[A]nd the Lord God formed man of the dust of the ground, and breathed into his nostrils the breath of life." How is one to decide which method was used by God to start life in Adam?

Clearly, creationists believe that the Bible is usually the most important book in their lives for knowledge acquisition, whereas evolutionists consider that material analyses are usually the most important means of gaining knowledge. We don't wish to voice an opinion in that matter except to suggest that if you have a theological question you will usually get answers that are more likely to be true from a holy man than from a scientist, but if you have a materialist question you will usually get answers that are more likely to be true from a scientist than from a holy man. Perhaps ethical rather than theological or materialist issues will prove to be the most important in the long run.

Notice also that, although science itself is agnostic, that does not mean that a scientist must be agnostic as well.

C. WHAT ARE THE KINDS OF RELIGIOUS BELIEFS REGARDING CREATIONISM?

We consider five main creationist beliefs, the first three of which are creationist in the sense that they accept God as the originator of everything. However, important differences lie in the details.

1. *Strict ("young-Earth") creationism.* The strict creationist believes in the literal meaning of the book of Genesis: that the universe and its living organisms were created by God about six thousand years ago in six twenty-four-hour days, and that the great (Noachian) flood killed all organisms other than the ones in the ark when the waters exceeded the height of the highest moun-

tain. When we use the word *creationism* in this book, strict creationism is the kind of creationism to which we are referring.

2. *Theism.* The theist believes that God is responsible for the universe and all its contents. He further believes that God listens to and hears our prayers and does intervene in this world to serve His ends. The theist may well be an evolutionist. As such, he does not worry about Genesis contradicting the evolutionary point of view. He treats the Bible as providing moral and esthetic truths, and that is sufficient to justify his faith. Most organized Christian religions hold the theistic belief. If you want a God that is personal, theism answers the need. Any conflicts between the Bible and evolution are resolved in favor of evolution, because there is no conflict if the biblical words are regarded as poetic parables—if, "in the beginning," six days can be longer than twenty-four hours each. In ordinary English one may speak of "the days of our youth," where "days" does not represent twenty-four-hour periods but stretches of many hundreds of hours. Moreover, if the time that elapsed until the sun was created was one "day," then there is no conflict between evolutionists and Genesis.

3. *Deism.* The third belief is that God is the originator of the universe but that God set forth the rules for evolution at the beginning of the universe. These rules have not changed. This God does not intervene in the world. Science is the process of discovering God's rules. Both deists and theists are creationists in the sense that they believe that God started everything, but we are not using that sense of *creationist* to define the creationists of this book. In this book the term *creationists* will mean the strict, literal creationists.

4. *Agnosticism.* The agnostic takes no position on the existence or nonexistence of God. Indeed, for him, God is an irrelevancy. There is no conflict with evolution with this belief, because the rules being sought are not God's rules but those of the material universe.

5. *Atheism.* The atheist is certain that God *does not exist,* although there can be no material proof for such a conclusion if one accepts the separation—the independence—of the four areas of knowledge given earlier. If one accepts that only perceptual phenomena are objects of exact knowledge, then it follows that we cannot *know* that God does not exist—since, by definition, God is a supernatural being. Being an atheist in this very strong sense is a position that can only be held "on faith." People will often call themselves atheists in a weaker sense that means they simply do not believe in the existence of God—without making any claims about certain knowledge of God's nonexistence.

Most scientists, if they are not agnostic (or atheistic), are deists. Indeed, most of the founding fathers of our country were deists, which was a popular philosophy at the time. Such scientists preserve a God in their life while doing research to reveal God's rules.

D. WHAT IS SCIENCE?

We use both deduction and induction in the empirical sciences. Recall that deduction is reasoning from the general to the particular and that induction is reasoning from the particular to the general. When testing hypotheses, we use induction—that is, we observe particular results in controlled situations and see if we can generalize to the hypothesis we wish to confirm. But we also use deduction to modify or generate a new hypothesis that we hope will fit our observations more precisely. Science is very limited. It seeks only materialistic explanations of materialistic phenomena. Supernatural intervention is not a permitted explanation in science. This doesn't mean that God didn't do it, whatever "it" is—only that, if God exists and if God did do it, science will fail to discover that fact as the correct answer. This loss is a small price to pay for having a knowledge system that provides good, solid measures of confidence with respect to those material questions that science can profitably examine.

Science is a method of examining materialist assumptions and propositions to discover other unknown facts that must also be true if the propositions are true, predicting that they will occur, and looking for those other unknown facts to see if they do occur. If any of those observations contradict the propositions, then something is wrong with at least one of the original assumptions and needs correcting. Curiously, if many tests are performed and the predicted happenings always occur, the proposition is still not proved. Note that no proposition can be proved, only disproved. (Perhaps the next new experiment will provide the long-missing disproof.) Nevertheless, if more and more predictions are always found to be correct, the confidence level in the proposition increases, and if it happens sufficiently often, the propositions may be given the status of a theory. *Theory* in this scientific sense is clearly not a mere guess.

E. MARK TWAIN AND SCIENCE

Mark Twain poked fun at science in the following delightful paragraph on the measured erosion of the Mississippi River. He mocks the perils of too-casual extrapolation from limited data. (Note that the "Cairo" that Twain refers to in his remarks is not the capital of Egypt but the town at the lower tip of Illinois where the Mississippi and Ohio Rivers meet.)

> In the space of one hundred and seventy six years the Lower Mississippi has shortened itself two hundred and forty-two miles. That is an average of a trifle over one mile and a third per year. Therefore, any calm person, who is not blind or idiotic, can see that in the Old Oölitic Silurian Period, just a million years ago next November, the Lower Mississippi was upwards of one million three hundred thousand miles long, and stuck out over the Gulf of Mexico like a fishing-rod. And by the same token any person can see that seven hundred and forty-two years from now the Lower Mississippi will be only a mile and three-quarters long, and Cairo and New Orleans will have joined their streets together and be plodding comfortably along under a single mayor and a mutual board of aldermen. There is something fascinating about science. One gets such wholesale returns of conjecture out of such a trifling investment of fact.

The trouble with many types of extrapolation is the assumption of *ceteris paribus* (a New Latin phrase for "all things being held the same"). Often changes in other factors can affect the trend that we are looking at. For example, Julian Simon and Paul Ehrlich made a famous wager that the inflation-adjusted price of any five commodity metals that Ehrlich could choose would be lower ten years into the future. Ehrlich selected metals that had a short-term trend of increasing price: copper, chromium, nickel, tin, and tungsten. But the prices of all five metals declined in the long run, just as Simon had predicted. Ehrlich had expected that the increasing population would outstrip the growth in supply of the metals and cause prices to rise. However, although the population *did* grow, changes in technology such as using plastic instead of copper for pipes in plumbing and using fiber-optic cables as a substitute for copper wire in communications caused the price of copper to fall instead of rise. The lesson here is that it is dangerous to assume that all things will remain as they are today and that short-term trends must continue indefinitely into the future.

F. OTHER TYPES OF INFORMATION

We need to distinguish among three types of informational words that are different from those that describe the categories given earlier.

1. *Definition:* a statement explaining what a word means, often telling you a larger class to which an object being defined belongs and then giving a feature that distinguishes the object from the other members of that class. For example, a bat is a mammal that flies. (Flying squirrels don't fly, only glide.) *Mammal* is the class to which the bat belongs, and the bat is distinguished from other mammals by its ability to fly. A word may be defined more precisely, but the idea remains the same. Definitions are accepted as legitimate givens, not to be argued over except to clarify the definition. They need not be of existing objects. For example, griffins are mythological creatures that are half lion and half eagle.

Creationists have argued that Darwinism is a religion. They usually argue this in the context of trying to support their position that, if Darwinism is taught in the science classroom, then creationism should also be taught in the same science classroom. Their logic appears to be as shown in the following syllogism.

Premise 1:   Darwinists are frequently overzealous in their promotion of Darwinism.

Premise 2:   Being overzealous is a common attribute of religious people.

Conclusion: Darwinism is a religion.

The syllogism is invalid in that it confuses the philosophy itself with those who believe in the philosophy (the fallacy of equivocation). It also confuses "some" with "all." The second premise needs to be revised and an additional premise (also not true) needs to be added just to get a valid structure.

Premise 1:   Darwinists are frequently overzealous in their promotion of Darwinism.

Premise 2:   All people who are overzealous are religious.

Conclusion 1: Darwinists are religious.

Premise 3:   If a people are religious, their philosophy is a religion.

Conclusion 2: Darwinism is a religion.

Note that the conclusion of the first syllogism is the first premise of the second syllogism. The two syllogisms are both logically valid in the sense that if their premises are true, the conclusions follow. But premises 2 and 3 are *not* true, and so the argument fails.

2. *Fact:* something that occurs (American monarch butterflies fly south to Mexico) or that is characteristic of existence (dogs

have four legs). Moreover, griffins are a fact (i.e., they exist), albeit only in Greek mythology. Facts are usually observations that may be used to support or contradict proposed explanations or that are of sufficient interest in their own right to provide an incentive to find explanations for them. It is important to recognize that facts may be in error (such as "griffins lived in Ancient Greece").

3. *Theory:* a plausible general explanation of a large number of facts. There are a number of words that may be used to indicate various degrees of belief or uncertainty in the explanation. These range from *guess, supposition, hypothesis, opinion, assumption, surmise, conjecture,* and *theory* all the way to *law.* We need only consider the last two alternatives: *theory* and *law.* A proposition that has been tested hundreds of thousands of times and always found to be uncontradicted may be called a *law.* An example is the first law of thermodynamics, which states that matter can be neither created nor destroyed. A proposition that has been tested thousands of times and always found to be uncontradicted may be called a *theory.* Examples are Einstein's theory of relativity and Newton's theory of gravity. Propositions that have such strong support are quite reasonably treated as if they were fact. This is the first example of what biologists mean when they say that evolution is a fact.

A second meaning of *theory* is also treated as fact. If there exist objects that fulfill a definition, then the definition describes a fact (it exists). Evolution is defined as noncyclical change over time. As the pattern and kinds of fossil organisms clearly change continuously from the Cambrian (545 to 495 M.Y.A.) through to our current time, evolution is a fact—one that stimulates a material search for its explanation.

But *theory* is a tricky word. A third meaning of *theory*, found in colloquial or ordinary usage, is that of a guess—a very different meaning from the meanings in the preceding two paragraphs as applied to evolution. Creationists often say, in a disparaging tone, that evolution is only a theory, meaning that it is only a guess. The creationists' syllogism goes like this:

Premise 1:   Evolution is a theory.

Premise 2:   A theory is only a guess.

Conclusion 1:   Therefore, evolution is nothing more than a guess (and thus is not a known fact).

This syllogism is valid but is bad logic. Its illogical result comes from *changing the meaning* of the word *theory* from the first premise (a well-substantiated scientific truth) to the second premise (a guess). This is another example of the fallacy of equivocation, which we first encountered in chapter 1. The proper syllogism is:

Premise 1:   Evolution is a theory.

Premise 2:   A theory is a well-supported explanation of many observations.

Conclusion 1:   Therefore, evolution is a well-supported explanation of many observations (and thus is both a theory and a fact, but not a guess).

A process that William Whewell named the "hypothetico-deductive" method is used to confirm or falsify a hypothesis. By this method, we first make observations of the world around us, we form a hypothesis to explain those observations, we make a prediction (deduction) about an observation that we should expect to see in given circumstances, we construct an experiment to create the given circumstances, and we observe whether the predicted event occurs. If the predicted event occurs, our confidence in the hypothesis increases, and we try to deduce other expected observations. If the expected observations do not occur, then we know we must modify our hypothesis, deduce new expected events, and construct new experiments to see whether the predicted events actually occur. We use inductive logic to form a theory (a general rule) from a large number of observations, but we can never be certain that our theory is correct. Scientists speak of having high confidence in a given hypothesis if it has been thoroughly tested, if it adequately explains the know facts, if it makes accurate predictions of future events, and if it has never been shown to be false by any experiment or fact.

G. EVOLUTION

Evolution is defined as noncyclic change over time. The noncyclic requirement eliminates cases like summer, fall, winter, spring, summer, fall, and so on, which are changes over time but are not examples of evolution. This definition applies rather generally, and ordinary people are accustomed to using the same meaning of the word that we do, as the following examples illustrate.

1. *Automobile evolution.* As a first example, consider the evolution of the automobile. It is clear that the automobile today is barely reminiscent of the horseless carriage from which it began. Automobiles, of course, have a designer. But the changes of automobile character states do not have the distributional characteristics of biological organisms. (See chapter 4, section O, "Non-Intelligent Design.") Robert Rowland took a lighthearted look at automobile evolution by examining a popular sports car in the 1960s, the MG, in his magazine article "The Evolution of the

MG" (1967). Despite Rowland's put-on treatment of the subject, the evolution of automobiles is a fact.

2. *Language evolution.* Language has also evolved. Modern Americans have a difficult time reading Chaucer's *Canterbury Tales* or sometimes even the more recent plays of Shakespeare. Thus language evolution is a fact.

3. *Geological evolution.* The planet's geology evolves. Plate tectonics slowly drives the Indian landmass under the Asian plate, lifting the latter so high that it produces Mount Everest, the highest peak on the planet. But over time the rains and glaciers tear down the mountains and we get, along with more rounded mountains, the wide deltas of the Mississippi and Amazon Rivers, made from the sand washed down from the uplifted mountains. Thus geological evolution is a fact.

4. *Biological evolution.* Biological organisms also evolve, in the commonly understood, everyday meaning of the word. Typical organisms of the Precambrian era (before 545 M.Y.A.) such as cyanobacteria are very different from those found in the Cambrian era (545 to 495 M.Y.A.), or from the sharks of the Devonian (417 to 354 M.Y.A.), or the dinosaurs of the Cretaceous (142 to 65 M.Y.A.), or humans (Cenozoic, 0 M.Y.A.). These examples of evolution are facts rigidly meeting the definition of evolution. Thus, since we can find examples of organisms changing noncyclically over time, biological evolution is a fact—our second example of what biologists mean when they say that evolution is a fact. The only logical way to deny the evolutionist's conclusion is by denying the Darwinist's definition of evolution, and redefining it. But that is not legitimate here, because we find biological evolution

to be a fact using our definition. Creationists cannot change our definition and then call us wrong. This is another good example of the fallacy of equivocation.

Of course, given that we have so many wonderful fossils and other related phenomena (e.g., similarities in embryonic development), any good scientist wants an explanation of how these things came about. Evolution is a relatively simple explanation for thousands upon thousands of such observations. The large numbers of tests passed is sufficient to justify the claim that evolution is a theory; it is, then, both a fact and a theory. To call it only a guess is to misrepresent what evolutionists have shown and to change the meaning of the word *theory* in midargument.

CHAPTER THREE

Some Simple Math and Statistics

This chapter discusses some techniques of science, and the criticisms that creationists have of those techniques.

A. MATHEMATICS AND STATISTICS IN THE FLOWER GARDEN

Science makes much of mathematics and statistics. Math and statistics are covered only briefly in this book. It will serve our needs to see some examples of the scientific process and how computations lie at the heart of our degree of certainty. No emphasis is placed on proving mathematical formulas, but the reader should get a clear understanding of the general mathematical process of showing a hypothesis to be supported or not by the data. Such demonstrations have different methods depending upon the hypothesis being tested, but it will be an important advance to your understanding of science if you master the idea behind scientific tests.

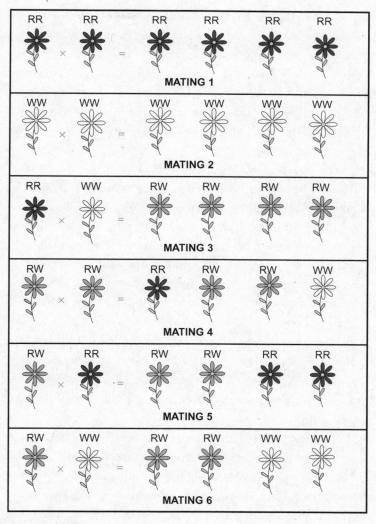

Figure 1. Flower matings.

Imagine you are growing flowers in your garden. You find that when plants with red-pigmented flowers are mated (crossed) with other red-flowered plants, they always produce red flowers (figure 1, mating 1), while similar matings between white-flowered plants always give white-flowered plants (mating 2). Being a curious person, you wonder what would happen if you mated a red-flowered plant with a white-flowered plant (mating 3), and your curiosity increases as you realize that there are at least two possibilities. The first possibility is that half the plants will be red and half white. The second possibility is that the flowers will all be pink (called *blending inheritance*), as if a white-colored "paint" diluted a red-colored "paint." You make the mating and discover—what? Guess the result before you read on to get the answer. (It is a good idea to have multiple hypotheses when doing science, because it helps you to think of experiments that might distinguish or predict different results between any two hypotheses. It also is good always to predict the outcome before doing the crucial experiment, because it develops your scientific intuition—your feeling about the way nature works. This is especially true when your prediction is wrong, because you then have to figure out where your reasoning might have gone awry.)

The correct answer in this case turns out to be: all pink. Don't worry if you guessed wrong. Depending upon how the genetics for this trait might have been set up, either possibility could have been correct. But now you think of another possibility: mating pink with pink (mating 4). Since mating red with red gives only red progeny, and mating white with white produces only white progeny, then surely mating pink with pink will give only pink progeny. But it doesn't. Half of the plants do have pink flowers, but a quarter of the plants have red flowers and another quarter have white flowers.

You now have enough information to propose a genetic mechanism that explains all these observations. It is assumed that, like most eukaryotic organisms, normal progeny get a complete set (pair) of chromosomes, receiving one chromosome from each parent. Moreover, there are other crosses that will test your mechanism. For example, what will you get if you mate a pink-flowered plant with one of the original red flowers (mating 5), or with one of the original white flowers (mating 6)? The answer for mating 5 is half red plants and half pink; the answer for mating 6 is half white plants and half pink plants.

Let us represent a red gene (or *allele*) by an R and a white gene by a W. Then the red-flowered plants will have two R genes, symbolized by RR. White-flowered plants will have two W genes, or WW. Assume that pink flowers have one of each kind of gene, or RW. Then the act of mating or crossing can be represented by the 2 × 2 matrices in figure 2. Write the genes in one of the flowers to be crossed down the left side of the matrix, and the genes in the other flower across the top of the matrix. Mating red with red (1) can only produce offspring having RR genes, which will be red like their parents. Similarly, mating white with white (2) can only yield white WW offspring. Mating red with white (3) can only produce offspring with RW genomes, and these are pink.

But things become more interesting when one mates two pink RW plants (4). Now the offspring have a 25 percent chance of inheriting an RR or a WW genome, but a 50 percent chance of the hybrid RW. Hence the result is one quarter each of red and white, and half pink. Finally, you can see from the matrices

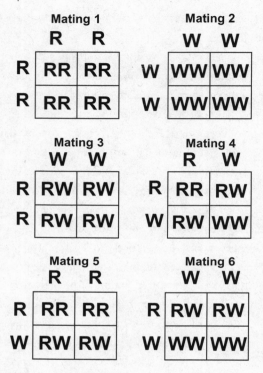

Figure 2. Allele diagram.

in figure 2 why mating red with pink (5) yields half red and half pink offspring, and mating white with pink (6) produces equal numbers of white and pink.

These two-lettered pairs of genes are called *genotypes*. The expressed result—in this case, the flower colors—is called the *phenotype*. When two genes, such as R and W, are not identical, they are referred to as *alleles*. The justification for the proposal that flowering plants carry two genes for a trait, and share their genes with offspring when mated, arises from the experimental observations with which you started.

But in a real experiment, some consideration of statistics is required. So far, simple reactions have been discussed—counting genes to find phenotype frequencies of one-quarter, one-half, or one in the plant population. In actual experiments, frequencies are not likely to be found to be those exact amounts. We can make an analogy to flipping a coin. For a fair coin flipped ten times, your statistical expectation is to see five heads and five tails. In actual practice, you may get a markedly different result—say, eight heads and two tails. Our statistical expectation of half heads and half tails will become more accurate in actual experience when we use a very large sample. If we flipped the coin 2 million times, we are likely to get results that are very close to half heads and half tails. In our flower example, if our random sampling of phenotypes is small, the results may be quite different from what our hypothesis predicts. Are the observed frequencies *close enough* to their expected values to be scientifically acceptable as support for the hypothesis? "How close is close enough?" is the question to be addressed at this point.

Suppose you wish to try one more test of your new genetics using one-quarter, one-half, and one-quarter as the expected (i.e., the most probable) frequencies of red-, pink-, and white-flowered plants, respectively. This involves matings among the pink plants growing in your garden, and your expectations are shown in mating 4 of figure 2. You go out and record the number of red-, pink-, and white-flowered plants in a random sample of 200 plants. Let us say that your experimental counts are 61, 98, and 41, respectively. Are those values *close enough* to the expected 50, 100, and 50, respectively, to regard them as not significantly different from expected values and thus to support your new genetics? Make your guess, and then read on.

There is an easily calculated statistical test called a *chi-square* that can be used here. The test measures the likelihood that the *null hypothesis* is valid or not. The null hypothesis is a statistical hypothesis that there is no difference between observed and expected data. What is called a *Type I error* occurs when we believe we are observing a difference when in fact there is none. That is, we reject the null hypothesis when we ought not to reject it. What is called a *Type II error* occurs when we believe there is no difference when in fact there is a difference. That is, we fail to reject the null hypothesis when we ought to reject it. The chi-square test is a purely mathematical test that has nothing to do with plants, or hybrids, or even evolution or any other branch of science per se. It gives, by lookup in a table, the probability of obtaining a frequency count that is as different from expected values as your observed frequencies are. We will not go into the mathematics here, but the chi-square test is a standard measure of likelihood of deviation from expected numbers. In our example, the 61, 98, 41 results have a probability of 0.045 of being compatible with the expected 50, 100, and 50.

	red	pink	white
Expected	50	100	50
Observed	61	98	41

The probability that the expected model is correct is 0.045, or 4.5 percent.

This means that there is only a 4.5 percent probability that a random sample of two hundred flowers will have sampled frequencies this different from the expected one-quarter, one-half, and one-quarter values. The most common cutoff point for rejecting a null hypothesis (the probability of obtaining the observed frequencies) is 0.05, or one in twenty. Any probability less than 0.05 is considered small enough to reject the hypothesis. Thus your hypothesis is rejected; your theory is invalid or is, at least, incomplete. Note that this cutoff point is arbitrary. The chi-square test shows that what we are observing is more likely due to chance than to the factors we are controlling for our experiment.

So what happened? A null hypothesis is the collection of assumptions necessary to calculate the expected values. If the test rejects your hypothesis, at least one of those assumptions must be wrong. Which one(s)? Further experiments are needed to reveal the answer. Perhaps the sample was not truly random? You can test that by taking a larger sample or making extra certain that your new sample of two hundred flowers has been chosen at random. Let us say that this second trial gives essentially the same result, once again rejecting your hypothesis. Then some other assumption may be wrong.

Until now, questions about the genotype frequencies (RR = red; RW = pink; WW = white) have been asked that involve thinking about two nonidentical copies (RW) of the same gene (alleles) together in the same individual parent plant. Consider the gene frequencies separately as R and W. There is nothing here that requires that the gene frequencies be equal. Suppose that the frequency of the red gene was 0.55 and that of the white gene 0.45. If we make that assumption, then the expected frequencies of the genotypes are as follows:

$$RR = 200 \times (0.55)^2 = 60.5$$
$$RW = 200 \times (0.55 \times 0.45) \times 2 = 99.0$$
$$WW = 200 \times (0.45)2 = 40.5$$

	RR	*RW*	*WW*
Expected	60.5	99.0	40.5
Observed	61	98	41

The probability that the expected model is correct is 0.90, or 90 percent.

By the chi-square test, there now is a 90 percent probability that your model is good. This time the hypothesis is not rejected, and your revised genetic mechanism looks promising, pending the results from other crucial tests yet to be made. It should be emphasized here that a statistical analysis merely defines our probability of being wrong. We cannot "prove" our hypothesis. However, we can build our confidence in our theory by demonstrating that our theory has a good correlation with observed data.

The process shows that you had a simple hypothesis: that all progeny had the color of their parents, and at first it looked good (matings 1 and 2). Then you found a case where the result contradicted your simple hypothesis because there could be more than one color for some parental types (mating 3). That led to a testable revision of your hypothesis (called *blending inheritance*). Further testing (matings 4, 5, and 6) did not support the revised version of your hypothesis, so you tried yet a third hypothesis: examining allele frequencies. This passed the statistical test, and thus your idea was tentatively accepted.

This illustrates the basic method of science:

1. Hypothesize.
2. Test.
3. Revise the hypothesis if necessary.

4. Test again.

5. Continue until you run out of different ways to test the hypothesis.

It must be like this to be a scientific method. And, only in science, the ultimate criterion of acceptable belief is *the predictability of the outcome of further tests*. This rejection of tentative hypotheses is called *falsifying*, proving that something is wrong (Popper 1959). *A theory or hypothesis that cannot be tested is not allowed in science; it is inherently unscientific precisely because it cannot be tested. And the most unscientific hypothesis of all is the simple assertion "God did it."*

B. HOW CREATIONISTS CALCULATE THE PROBABILITY OF GENERATING A PROTEIN SUCH AS CYTOCHROME C BY A RANDOM PROCESS

What is the probability of assembling the mitochondrial protein cytochrome c by randomly connecting one amino acid at a time to the growing chain? The protein is 104 amino acids long. (For simplicity we'll say 100, which is conservative.) The equine cytochrome c sequence begins with G-D-V-E-K-..., where each different capital letter represents a different amino acid. In this example, G = glycine, D = aspartate, V = valine, E = glutamate, and K = lysine. (It isn't necessary to know the names of the amino acids—only that there are twenty of them, one for each of the twenty different amino acids encoded in the DNA.) Assume that there are equal (but *very* large, effectively infinite) amounts of each amino acid available. They are all well mixed in a very large bag into which you cannot see. This will mean that the probability of your drawing any particular letter (amino acid) is one part in twenty, or 0.05. What is the probability that G is selected on the first draw? It should be obvious that the probability is 0.05, the fraction of the total for each amino acid. So far so good. But what, then, is the probability that our second draw will be the required D? Again it is, of course, 0.05. Now comes a simple complication. What is the probability of getting a G on the first random draw *and* a D on the second random draw? Since the probability is not changed for the different draws, the joint probability is $0.05 \times 0.05 = 0.0025$, or one chance in four hundred (which is $1/20 \times 1/20$). As you can see, the odds are very much against your randomly sampling even two of the amino acids in the correct order: first G, then D.

Of course, if you have a lot of time, enough for several hundred trials, you can expect to get your G-D beginning occasionally. But now you need to sample randomly for the third amino acid, valine, represented by the letter V. The probability of getting that is still 0.05, but the probability of getting all three in that order is $(0.05)^3 = 0.000125$ or $1/8,000$, and you will need a lot more time to get a correct sequence.

By now you should realize that the probability of randomly getting the E after the first three letters in proper order is $(0.05)^4 = 0.00000625 = 1/160,000$. Moreover, that means that the probability of your getting the fifth letter in proper order, K, is $(0.05)^5 = 0.000000313 = 1/3,200,000$. The random probability of getting all one hundred amino acids in the proper prescribed order is $(0.05)^{100} \cong 10^{-130}$. And this number is substantially smaller than the reciprocal of the total number of atoms in the universe. The number is so small that most statisticians would readily conclude that the chances of the event happening are, for all practical purposes, virtually indistinguishable from zero.

So what have you (and the creationists) proved? You (and they) have proved that the probability of getting a string of one hundred specified amino acids in proper order by a mechanism of randomly picking the correct amino acids from a large pool and adding them onto the growing right (carboxyl) end of the chain is so small that no protein that size or greater could ever have been produced. And you have examined only one of many proteins in the living cell that need to be constructed. This is one of the reasons given by the creationists for their assertion that life could not, without God's intervention, have originated on Earth—nor anywhere else in the universe.

But the reasoning is in error. What has been proved is that no protein of one hundred or more amino acids could ever have been formed *by this particular random mechanism of protein synthesis.* The creationist model requires that a specific one of the twenty amino acids be randomly obtained and incorporated into the growing protein. The process is repeated another ninety-nine times to get a protein one hundred amino acids long.

Molecular evolutionists have discovered that there is a messenger RNA chain that tells the protein construction machine (ribosome) what the next amino acid should be (say, a D). If the next sampled amino acid is not a D, the machine rejects that amino acid and searches for another one. It repeats the process until a D is found, adds the D to the growing chain, and then goes back to the messenger RNA to find out what the next amino acid after D should be. In this way the probability of getting the correct amino acid at each individual step jumps from 0.05 to

1.00, and only occasional errors occur during synthesis of millions of copies of each of hundreds of different proteins. The correct probability is not 10^{-130} (couldn't happen) but 1.00 (bound to happen). An analogy can be made to the attempt to draw a royal flush poker hand from a standard deck of cards. The odds of five randomly drawn cards forming a royal flush are only 1 in 649,740. But if you are allowed to repeatedly draw and discard back to the bottom of the deck selections until you come to each particular card that you wish to retain, then you are certain to be successful. The protein-construction machine works like that.

A word about the use of the word *conservative* in the statistical statements in this book. Consider again the equine cytochrome c. A sequence of 100 amino acids was used for the protein length when in fact the sequence is 104 amino acids long. What kind of error is associated with that difference? In fact, if the correct length of the sequence, 104 amino acids, had been used, the estimated probability would have been even lower, by a factor of $(0.05)^4 = 0.00000625$, or an overall probability of 10^{-135} instead of 10^{-130}— five orders of magnitude (five powers of ten) smaller. The creationists want a significantly low probability so they can say that the proposed random protein-assembling mechanism cannot be true. Had the correct length been used—104 amino acids—it would have yielded an even lower probability than the already convincing (im)probability that you just calculated. Thus the actual calculation that you made is biased against your getting the result that the creationist desires. But despite the bias against calculating a low probability, you still obtained an awesomely significant low probability. Thus the original calculation is conservative because the true probability is even lower than the value actually calculated. Similarly, using all amino acids at equal frequencies instead of at their known frequencies gives us a greater probability but one that is still very small, thus enabling us to reject the hypothesis that equal amino acid frequencies is the correct model. And again the calculation is conservative because the computed p (probability) value, although it is greater than the true value, is easily small enough to reject the creationist model.

C. THE AGE OF THE EARTH

Now let us turn to a key problem in thinking about how our world came about: the age of the Earth itself. Information from decaying radioactive molecules has been used to date ancient events, so it is appropriate to examine how this method works. This section, along with sections D (potassium/argon dating of fossils), E (radiocarbon dating of fossils), and G (Occam's razor and the age of the Earth), can be skipped if you are comfortable with the method of dating given by the various approaches that permit dating of ancient events. There are many methods for dating fossils. This discussion will detail only two revealing comparisons of the complete set of isotopes possibly related to the natural earth, plus the $^{40}K/^{40}Ar$ (i.e., potassium/argon) ratio method. It is important to recognize that when the results of various methods of dating are properly compared, *the same dates are obtained*. For more details about this topic, see chapter 3 in Kenneth Miller's *Finding Darwin's God* (1999), which is particularly lucid.

The periodic table of the elements contains thirty-four naturally occurring radioactive isotopes that might be found on Earth.

Isotopes are atoms of the same element (i.e., they have the same number of protons) but with different numbers of neutrons (atomic weight). Each kind of radioactive atom has a different half-life. A half-life is the time that normally passes in order for half of the radioactive atoms to decay. Figures 3a and 3b illustrate the exponential character of this process. If you start with half of 64,000, or 32,000, radioactive atoms, then after a second year there will be half of 32,000, or 16,000, radioactive atoms left to decay; after a third half-life there will be 8,000 atoms left; after a fourth, 4,000; then 2,000; then 1,000; and so on. If the number of radioactive atoms remaining is plotted against the number of half-lives of decay, you will get a curved exponential decay line (figure 3a); but, if plotted on logarithmic paper, you will get a straight line (figure 3b). From this, if you know the half-life of an element and can calculate how much of the original sample has decayed, then you can calculate, or read from the graph, the age of the sample.

So when was the Earth formed? Anything close to zero years ago means the Earth was formed very recently. Several "young-Earth" creationists (e.g., Dr. Terry Mortenson) say about six thousand years ago. If the Earth was formed an infinite number of years ago, that means it has always existed. Perhaps its true age is somewhere in between these limits, and maybe the decay of radioactive elements can tell us the answer.

Those radioactive atoms that decay most rapidly—those with extremely short half-lives—would have decayed effectively to nothingness, and thus would no longer be a significant presence on Earth. Of the thirty-four potentially observable radioactive isotopes we have looked for on our planet, eighteen have been detected. Eleven have never been detected, and another five are disregarded here because they are created anew from other

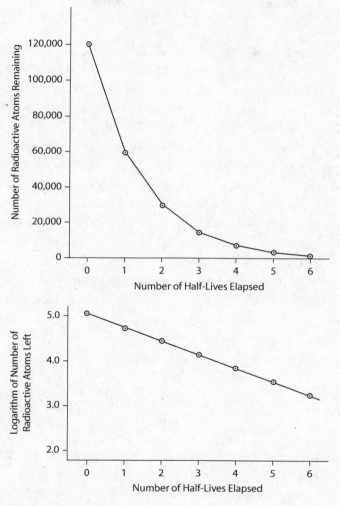

Figures 3a and 3b. Radioactive decay.

radioactive atoms, making them undependable for the dating process.

What are the half-lives of these eighteen useful isotopes? All of them have half-lives *greater* than 8×10^7 years (80 million years). Moreover, all eleven isotopes that are no longer present on Earth except when man synthesizes them have half-lives *less* than 8×10^7 years. This remarkable dividing point means that the Earth has existed for enough half-lives that no detectable traces of those atoms remain. How many half-lives is that? It depends upon the sensitivity of the detection apparatus (the Geiger counter), but twenty half-lives represents a sensitivity to detection of about one part per million—a reasonable value, although only an approximate one. That would mean the Earth was formed about $20 \times 8 \times 10^7 = 1,600,000,000$, or 1.6 billion, years ago. If the error margins were plus and minus two half-lives, then the range of the estimated age of the Earth would be from 400,000,000 (400 million) to 6,400,000,000 (6.4 billion) years ago—numbers that bracket nicely our best present date for the Earth's origin: 4,500,000,000 (4.5 billion) years ago.

D. THE DATING OF FOSSILS: POTASSIUM/ ARGON DATING

A common radiometric method is potassium/argon dating ($^{40}K/$ ^{40}Ar). ^{40}K is potassium-40 (19 protons + 21 neutrons). It is radioactive and, through the loss of one proton and one electron and the gain of one neutron, becomes ^{40}Ar (18 protons + 22 neutrons). In a volcanic lava flow the rock is liquid, and dissolved gases such as ^{40}Ar can escape. Thus when the molten lava first cools and solidifies it contains no Argon gas, only ^{40}K. But since ^{40}K is radioactive, its decay product, ^{40}Ar, begins to accumulate because it cannot escape from the now-solid rock. Because the total of ^{40}Ar formed from ^{40}K plus the ^{40}K remaining equals the initial amount of ^{40}K present, and because we know that the half-life of ^{40}K is 1.3 billion years, we can calculate the number of half-lives that have elapsed since the rock solidified. So we can tell how long ago the volcano erupted. (Note that radioactive half-lives are not affected by changes in temperature.)

Because molten lava totally destroys any organisms in it, the method must be modified to give fossil-dating information. After the lava cools, sediment may accumulate above the rock layer. If an organism dies and falls on top of the sediment, it may be fossilized there as more sediment falls. Whatever the fossil's age, it cannot be more than the age of the lava flow below it. If a second lava flow occurs above the fossil, this lava flow can also be dated, and its age cannot be greater than that of the fossils below it. Simply stated, the age of the fossil must lie between the ages of the two lava flows, one above and one below. That permits us to interpolate between the two dated lava flows to get an estimate of the fossil's age. There are many other radiometric dating techniques, including rubidium-strontium and uranium-lead. Each radioactive decay process has a different range of time over which it is useful. Zircon crystals can survive long geologic processes (like erosion), so they are useful for uranium-lead, fission-track, and helium techniques of dating.

E. THE DATING OF FOSSILS:
RADIOCARBON DATING

The Earth's atmosphere has a particular ratio of carbon-12 to carbon-14. Carbon-12 is the naturally occurring form of carbon. The cosmic rays that enter our atmosphere can react with ordinary nitrogen to form small amounts of a substance known as carbon-14. Both the carbon-12 and carbon-14 forms of carbon are dispersed throughout the food chain by being absorbed in the form of carbon dioxide, and by the consumption of plants or animals that have previously absorbed both forms of carbon. As long as an organism is alive, there is constant metabolic exchange of both forms of carbon between the organism and its surroundings. The organism itself will have a fixed ratio of carbon-12 to carbon-14 contained in its body that mimics the ratio found in the atmosphere. But once the organism dies, it ceases to take in either form of carbon. The carbon-12 variety is stable, and the amount of it contained in the organism at the time of death remains fixed. However, the carbon-14 variety of carbon is radioactive and unstable. Carbon-14 will start to disappear from the body of the organism at a "half-life" rate of 5,730 years (half the existing amount will be transformed into nitrogen every 5,730 years). Thus, the longer that organism has been dead, the more carbon-14 will disappear without being replenished. Since we know the half-life rate at which the carbon-14 decays, we can calculate how much time has passed since the organism died. Radiocarbon dating is limited to specimens that are about fifty thousand years old and that are organic (a fossil of an organism that has metabolized carbon during its lifetime).

Needless to say, one has to be careful in carrying out experiments in radiocarbon dating (or indeed in any other scientific process). Any one or more of the following scenarios are possible:

1. If an organism is found in water, it could have a different carbon-12 to carbon-14 ratio than the atmosphere because of the differing solubility of carbon dioxide in different

bodies of water. Plants growing in water can be affected this way, as well as aquatic animals such as seals and whales.

2. Very recent samples may not have experienced enough carbon-14 decay for the decay to be measurable.

3. Samples older that about fifty thousand years may have effectively run out of radiocarbon to measure, and so the computation will be inaccurate. Paint cannot be dated this way because the oil in paint is ancient carbon.

4. Most fossils cannot be dated because of contamination—as, for example, by the use of solvents such as acetone to clean up the fossil.

5. The method assumes that the rate at which carbon-14 is produced in the atmosphere is known and constant. It turns out that it is not constant. However, the amount by which it changes is determinable and thus can be compensated for. It does mean, though, that some of the dates determined a few decades ago are inaccurate and need to be reanalyzed.

Very dependable dates have been obtained since about 1970 by adhering strictly to analytical procedures. Verification by using more than one method and more than one sample is desirable.

F. OTHER METHODS OF DATING

Four other useful dating methods are as follows:

1. Dendrochronology: Also known as "tree ring" dating, this method was developed in 1914 by Andrew Ellicott Douglass. It makes use of the fact that the size of tree

rings varies from year to year in the particular region where the tree grows. The amount of a tree's growth each season sets up a pattern of thin and thick rings that can be recognized. We can cross-reference several overlapping patterns of rings, including those of living trees, to allow us to count backward in time. A master sequence can then be constructed consisting of all the available overlapping patterns. Thus, a sample can be dated according to where its pattern fits into the master pattern. Dendrochronology is often accurate to within a single year, but it is limited in the range that it can cover (about eleven thousand years for northern Europe). Another problem is that for older dates, the system depends on the longer sequence of the bristlecone pine, which is rarely found among artifacts that require dating. Two dating methods similar to dendrochronology are measurements of layers of coral reefs and measurements of varves (layers of sediment).

2. Amino acid racemization: This technique makes use of the fact that amino acids occur in two forms that are chemically alike but differ in the same way as a right-hand glove differs from its left-hand mate (mirror images, called *isomers*). These amino acid forms consist of the "D" type (from the Latin *dexter:* "right-hand") and the "L" type (from the Latin *laevus:* "left-hand"). Amino acids in living organisms are 100 percent "L" type. After an organism dies, its amino acids slowly convert over to the "D" type in a process known as *racemization*. In a fossil containing amino acids, the greater the percentage of "D" type detected, the older the fossil is. This

method can date fossils with amino acids in them to as far back as about one hundred thousand years ago. One difficulty with this dating method is that the rate of racemization is potentially unsteady. It can differ according to the type of amino acid and such factors as temperature and soil acidity. Due to the factors that can alter the rate of racemization, some scientists consider the amino acid racemization method to be a relative dating method rather than a chronological dating method.

3. Geomagnetic reversals: Iron minerals in some types of rock and clay can become magnetized when subjected to high heat (e.g., volcanic activity). The magnetized particles align themselves with the Earth's magnetic field, like the needle of a compass. After cooling, the aligned particles leave a record of the direction of the Earth's magnetic field at the time when the sample was formed. The direction of magnetic north is known to have changed over time and also to have reversed polarity numerous times (about four hundred known reversals). Thus, examination of the alignment of the particles gives us information about the orientation of the magnetic poles at that time. Nearby fossils can then be approximately dated as occurring between subsequent realignments of the Earth's magnetic field.

4. Fission tracks: Fission is the process by which an atomic nucleus splits into parts. When a fission fragment passes through a solid, it leaves a *track*—a path of detectible damage in the direction through which it has passed. Uranium-238 is most useful for this purpose because it

undergoes spontaneous fission and will produce a significant number of fission tracks over time. Glass and zircon are usually the materials that are examined for fission tracks because these materials are suitable owing to their abundance, track retention, and presence of uranium. By counting the tracks, and taking into account the amount of uranium present in the sample and the rate at which uranium decays, we can determine the age of the sample. A disadvantage of this method is that unless the sample is in the range of about one hundred thousand years old or older, we will not find a significant number of fission tracks for the purpose of dating.

And there are still quite a few other dating methods. When more than one method is applicable for the age being determined and great analytical care is taken, they agree with each other (within the margin of error). Thus the calculations are reliable and repeatable, providing useful dates.

G. OCCAM'S RAZOR AND THE AGE OF THE EARTH

Let us first look at two different ways of estimating the age of the Earth and then see how to use Occam's razor to judge which method is the better. A sample of very old rock has v units (concentration) of a specific radioactive element and z units of its product. Then the fraction of the original radioactive element that has still not decayed, or f, is represented by the formula $f = v/(v + z)$. The number of half-lives during which the radioactive element has been decaying is $h = (ln\,f)/H$; where

$H = (ln\ 0.5) = -0.693$ (where "*ln*"is the natural logarithm). Then the number of years (ago) that the radioactivity came to reside in your sample is $y = Dh$ where D is the number of years of decay per half-life. Expressed mathematically, three simple equations are obtained (capital letters represent constants):

$f = v/(v + z)$ fraction of the original isotope that remains

$h = (ln\ f)/H$ number of half-lives, where H is $ln(0.5) =$
                 -0.693

$y = Dh$ number of years the isotope has been decaying,
                 where D is the number of years per half-life
                 of the isotope (and depends on the particular
                 radioactive isotope being studied)

For example, if v and z are 0.708 and 1.292 units, respectively, then $f = 0.354$. So the number of half-lives during which the decay process has been going on is $h = (ln\ 0.354)/-0.693 = 1.5$ half-lives. Finally, $y = 1.5 \times D$ years, where D is the half-life of the radioactivity. If, in this hypothetical example, D is 3 billion years per half-life, then the sample would be 4.5 billion years old. What could be simpler?

"Young-Earth" creationists who believe that the Earth— indeed, the entire universe—is only six thousand years old must use a different approach. Believing that the Bible is always literally true irrespective of anything that might contradict it, the creationist begins by assuming that *y must* equal 6,000—and that therefore, the isotope half-life used by the evolutionist must be incorrect somehow, and that its correct value can be calculated from the same three equations. If the fraction *f* is taken to be 0.354 and *h* is taken to be 1.5 (as before), then *D* must be 4,000 years per half-life in order that *y* = 6,000 years.

The principle of *Occam's razor* gives us a rule to help decide which of two solutions to prefer. The English philosopher and theologian William of Occam (1285–1349) laid down the principle that "[e]ntities should not be multiplied needlessly." This rule is interpreted to mean that the simplest of two or more competing solutions is preferable and that an explanation for a newly discovered phenomenon should first be attempted in terms of what is already known. Let us see how it works in the two solutions to the age of the Earth: solution s_1 ($y = 4.54$ billion years) and s_2 ($y = 6{,}000$ years).

For s_1, if one accepts the previous values of f and h, then a value of $D = 3$ billion years per half-life will imply that $y = 4.54$ billion years. In this case, D comes from observed data, and y is determined from the third equation ($y = Dh$).

For s_2, one also accepts f and h as before. The Bible is used to determine that y is 6,000 years, and y is used to determine the value of D.

For s_2, the calculation requires an additional entity: some mechanism to explain why a value of $D = 4{,}000$ years per half-life differs from the observed data. Thus s_2 is rejected in favor of s_1 because s_1 requires fewer entities. It may be said to be simpler, to require fewer assumptions; that is, it is more parsimonious.

One explanation suggested is that the data fit solution s_1 perfectly, giving a date of 4.54 billion years ago, but the true date of origin is nevertheless six thousand years ago. It is suggested that *God has changed the isotope ratios* so the true date of origin is not obtained by using the three equations. There are four responses that the evolutionist should make to this train of reasoning:

Response no. 1 is that the creationist is forced to make more assumptions, including that God is responsible and that He can fix each of the isotopes' frequency to whatever value He desires. Since the creationist has more assumptions, then by Occam's razor, the evolutionist's solution is better.

Response no. 2 is that the evolutionist is using theory and data to discover when the Earth was created, whereas the creationist "knows" the answer he wants and makes the assumptions so as to get the "right" answer of six thousand years. It is bad science to assume the answer you want to be the correct one and then to work backward from it. An "ad hoc hypothesis" is a hypothesis specifically created in order to prevent a theory from being falsified or discredited. A famous example is the use of "epicycles" to explain the motion of planets. When the Ptolemaic theory of other planets moving about a circle centered on the Earth was found to be inaccurate, "epicycles" were introduced. The other planets were said to move in a smaller circle (an "epicycle") that centered on a point that moved about a larger circle centered on the Earth. In this way, observed errors could be accounted for and the theory of Earth-centered planetary motion could be retained. But as observations became more accurate, as many as forty "epicycles" were needed in some cases to explain the observations! The principle of Occam's razor tells us that the simpler explanation is probably the better one. The heliocentric (sun centered) system favored by Copernicus explained the same observations in a much simpler manner and thus came to be accepted in place of the Ptolemaic system. Please keep in mind that it is perfectly legitimate to revise your original theory and to come up with explanations of why you think the observations do not fit your original theory. But there may come a point where your original theory has so many "patches" that it is better to abandon the theory for a simpler explanation.

Response no. 3 is that invoking God is not a scientist's option while practicing science; supernatural involvement is not allowed in science.

Response no. 4 is that if God changed the frequencies of radioactive isotopes to make the newly created radioactive elements *appear* very old—that is, although they seem old, they are only six thousand years old—*God has deliberately misled us.* But why would God do such a thing? It contradicts His commandment that you should not bear false witness. It seems unlikely to us that a benevolent God would deliberately deceive us. This approach is reminiscent of that of an early-nineteenth-century theologian, Philip Gosse. In his book entitled *Omphalos* (Greek for "belly button"), Gosse explained away the fossil record by stating that God Himself had laid down the fossils in a way that appeared to suggest an ancient Earth, in order to test the faith of the faithful! Gosse's friend the minister Charles Kingsley was horrified by the *Omphalos* theory. Kingsley felt that a newly created Adam's navel or recently formed fossil record would make God tell a lie—thereby making Him "God-the-Sometime-Deceiver" (see Hardin 1980).

Note that in comparing two explanations—one creationist, the other evolutionist—the simpler of the two is normally the evolutionist explanation, except when the creationist explanation is "God did it." Thus, by Occam's razor, the evolutionist's explanation is to be preferred. Note further that Occam's rule originated long before Darwin came along and has nothing to do with evolution. Thus, there is nothing special about the rule that would cause it to be biased in favor of evolution.

H. THE RARITY OF FOSSILS

For every organism that dies and leaves behind a fossil that is subsequently discovered, millions of other organisms leave behind no trace of their prior existence. Look at the plot in figure 4. Each row is a new "generation" for the population of organisms whose size is arbitrarily held constant at sixteen individuals (row width). Nodes represent an individual that either has progeny (a dot at that node) or has none so that lineage goes extinct (an × at that node). The highlighted pathways mark the ancestry of those few organisms that made it to the present day. Isolated fossils are represented by the nodes labeled C, D, and E. Node B is the most recent common ancestor of C and D. Node A is the *ceneancestor*—the most recent common ancestor of all the taxa (fossils) being examined. Note how few fossils

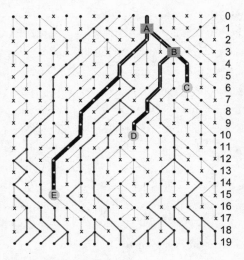

Common ancestors
"A" is cenancestor:
most recent common
ancestor

Isolated Fossil

Organisms alive
at a given time

Figure 4. Fossil phylogeny.

have been recovered, although each node represents an organism that once existed. Note especially that none of the isolated fossils have living progeny (row 19). This genealogical representation would be quantitatively more accurate if each dimension were multiplied by one thousand, but still, only the three fossils (C, D, and E) are found. This is so commonly the case that it is often said that a rock-hunting fossil gatherer (paleontologist) never discovers a direct ancestor of a currently living organism (the paleontologist looks for "collateral ancestors"). An obvious though rare exception occurs when the fossil had been a pregnant female and one can thus compare the parent and child.

I. SEQUENCES AND TREES

It is sometimes useful to draw diagrams showing a nested relationship of a group of organisms descending from a most recent common ancestor (cenancestor) together with the cenancestor itself. Such analysis is called *cladistics,* which focuses on the shared derived characters. Organisms have traditionally been grouped according to their appearance or common behavioral traits, but such classification poses problems, because we cannot be sure which of the features that we observe are the most relevant. In more recent years we have another tool to help us classify organisms: we can use our knowledge of the base sequences of DNA. The bones and teeth of horse and cow are more similar to each other than either set of bones and teeth is to the bones and teeth of mice. This suggests that the cenancestor of horse and cow, if it still existed to be studied, would be more like the modern species than is the common cenancestor of mouse, horse, and cow. Said differently, the horse-cow cenancestor is the most recent of all three pairs of taxa. Other methods also support that inference.

An extremely powerful method is the comparison of amino acid sequences of proteins, or base sequences of DNA, among the species in question. Observed differences in sequences between organisms can be used to construct an evolutionary tree, and the number of trees that can be constructed for a given set of organisms is limited only by the number of different proteins or different DNA segments shared by all three of the organisms— essentially infinite, for all practical purposes. One of the most convincing aspects of this sequence comparison and tree construction is that the results tend to be the same no matter what protein or what DNA region is chosen.

Consider the three (hypothetical) DNA sequences shown at the top of figure 5. The sequences are labeled C, H, and M for a hypothetical cow, horse, and mouse, respectively. A small dot over the letters indicates DNA nucleotides that differ from one sequence to another. This is an optimal alignment, because no other alignment gives fewer "dotted" nucleotides. (The alignment is obvious here, but in real cases of more differing sequences, computer programs can help search out the optimal alignment.) For every pair of sequences, count the number of DNA differences between them. These differences are displayed in the little three-by-three matrix to the right of the sequences. In this model system, cow and horse differ at eight positions, horse and mouse at eleven positions, and cow and mouse at thirteen.

Now consider how far the three species differ from their nearest common ancestor. In the circle diagram of figure 5, let c be the number of DNA differences between cow and the

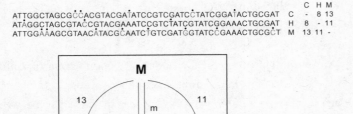

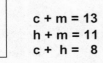

Figure 5. Computing branch lengths.

(unknown) common ancestor, let *h* be the number of differences between the horse its same ancestor, and *m* that for the mouse. Of course we have no sequence information for the common ancestor, but we do know from the data in the difference matrix that $c + h = 8$, $h + m = 11$, and $c + m = 13$. These numbers represent the relative evolutionary distances between the three species.

The conclusion is that the cow and horse are closer relatives (only eight differences) than either of the two comparisons (cow–mouse, thirteen differences; horse–mouse, eleven differences). This is what one would expect if Darwinism were true, because the teeth and bone information gave the same closest pairing. These are only hypothetical data drawn up to illustrate the method. But extensive real sequence data exist and have been analyzed for alpha and beta hemoglobins, for cytochrome c and many other proteins, and, more recently, for DNA

sequences. As long as the sequences are long enough, or sufficiently varied (i.e., containing enough "dotted" nucleotides), real
data always support the cow–horse relationship. When this happens many times and without exception, the relationship between
species is said to be well supported. When tests like this are performed many times, when the predictions of evolution continue
to corroborate the relationship of cow and horse, in preference
to the mouse (and similarly for other triplets of organisms), evolutionists properly conclude that the bases for those predictions
constitute a theory.

It would be misleading not to note that there are exceptions.
In the early days of the primate comparison among humans,
chimpanzees and gorillas frequently gave contradictory conclusions for different gene products. The problem disappeared when
longer sequences (around two thousand nucleotides) were used.
What was shown was that humans and chimpanzees were more
closely related to each other than either was to the gorilla. So
the problem was not that the primates were not relatives of each
other, nor that the methods used in the analysis were deficient,
nor that the relationships are different from gene to gene (although
there is awareness of such); it was simply that there was too little
information.

Note how that can come to be. Consider a pair of dice where
each die has six sides with dots from one to six on them. It is possible to load the dice so that when they are rolled the first die will
come up as "1" each time and the second die will come up as "6"
each time, so that the sum of the dots on the two face-up sides
is $(1 + 6 = 7)$ seven. You can test whether the dice are loaded by
rolling them many times and counting how often they give a 1
and a 6. With six faces on a die, the probability of getting any

one number on a single unloaded die is 1/6. The probability of getting a 1 on the first die and a 6 on the second is $(1/6)^2 = 1/36$. The probability of getting a 6 on the first die and a 1 on the second is $(1/6)^2 = 1/36$. The probability of getting either one of the above two cases is twice as great: $2 \times (1/36) = 1/18$. Therefore, if they are "fair" dice, you should see a 1 and a 6 with a frequency of 1/18 ($\sim$0.0555). So if you rolled the pair one hundred times you should expect we see a 1 and a 6 about six times. If you should see that combination twenty-seven times in one hundred rolls, you should be very suspicious. But notice that you cannot roll the dice only once, get a one and a six, and then declare that the dice are loaded. Similarly, only a few "dotted" nucleotides in the sequences may be too few (like one roll of the dice) to declare that chimpanzees are our nearest living relatives, even though the declaration is true.

J. TREE RECONSTRUCTION

In phylogenetic analysis, the most plausible tree that can be constructed from the data is the one that requires the fewest evolutionary changes in character. This is the *principle of parsimony*. Since evolutionary change is improbable, the principle of parsimony gives us the tree with the fewest required character changes. Thus it is the most likely tree. Although we cannot be sure that we know the "true" tree, the most parsimonious tree is the one most likely to be correct. Once again, we are invoking the principle of Occam's razor in deductive reasoning—that entities should not be multiplied needlessly. This does not guarantee that we have the correct solution—only that we have the most likely solution in the absence of other knowledge. Let us now take a look at how we can construct the simplest or most parsimonious tree (in terms of evolutionary changes required).

The number of differences between pairs of sequences for our cow–horse–mouse example is shown diagrammatically as distances around the circumference of the circle in figure 5. It is not treelike (although it is sometimes called an *unrooted tree*) and provides no sense of a parent and a child, of an ancestor and a descendent. Such a tree would require that in the interior of the circle are three branches (like the spokes of a wheel), with lowercase letters *c, h,* and *m* representing the unknown lengths of the three branches of the tree. Those lengths can be calculated, because there are three equations and three unknowns (bottom right of figure 5). Solving these three simultaneous equations yields $c = 5$, $h = 3$, and $m = 8$.

These three numbers can be used to construct a tree (see figure 6) showing the evolutionary relationships of cow, horse, and mouse. The tree indicates that, after the departure of the animals' genomes from a common ancestor, the mouse genome underwent six mutations down to the present day. The common ancestor of cow and horse underwent two mutations before the line branched. Since then, cow has experienced thee more mutations down to the present day, while horse has had five. Note that all their lines have different rates of evolution, although the differences are so divergent as to make rate studies invalid.

In short, protein and DNA sequence comparisons between species can allow one to deduce the "family trees" of the species and their ancestors. There will be differences in rates of accumulation of mutations depending on the organism and its environment. In this purely hypothetical example, if horses and cows still exist at the present day, it would appear that something

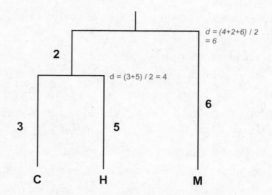

Figure 6. Phylogenentic tree (relative rate test). Distance (d) is in nucleotide substitutions or amino acid replacements.

about the horse caused it to accumulate mutations faster than the cow. But the connections between species, and their derivation from common ancestors, are unimpeachable.

K. IS THERE A MOLECULAR CLOCK?

The possibility exists that the rate of mutations fixed during evolution is sufficiently steady over time that one may estimate *when* the common ancestor of a pair of sequences occurred. The first task is to find whether the rate of evolution for the organisms under study has remained sufficiently constant during their divergence. As we just noted, in figure 6 the cow and horse have accumulated different numbers of mutations since their lines diverged. If the branch tips can be asserted to have been "observed" at the same time (a hundred years difference is most unlikely to have any effect because of the slow rate of evolution, excluding some RNA viruses), then the time from the branch tip back to their common ancestor must be the same, even if we don't know what it is.

If there is such a "molecular clock," and the accumulation of mutations in a given protein or DNA gene occurs at a relatively constant rate, then the numbers of mutations from the branch point of two species to the present day should be the same for both lineages. Are they? Clearly not. But are these differences statistically significant? No, in this example they are not. So the null hypothesis that the evolution of nucleotides is clocklike cannot be rejected in this hypothetical case. However, if one looks at large quantities of real sequence data, the clock hypothesis is almost always universally rejected. The clock is sloppy; it may not have a second hand, or even a minute hand, but it does have an hour hand. A given protein does not evolve at a constant rate throughout its evolution; some proteins, by their very nature, do accumulate mutations more slowly, or more rapidly, than others.

But that is not the end of the scientific value of a molecular clock consideration. It first requires that you be introduced to one or more statistical notes. Rates (and many other variables) often come out in the form $r \pm s$ where s is the 95 percent confidence interval. Thus a value such as 57.3 ± 8.7 is to be read as: "The true value of r lies, with 95 percent confidence, between 48.6 ($= 57.3 - 8.7$) and 66.0 ($= 57.3 + 8.7$). This is hardly the precision one would like, but for many cases, knowing that a value lies between 48.6 and 66.0 may be sufficiently accurate to rule out some hypothesis. Because it is a 95 percent confidence interval, stating the conclusion that r lies within the interval means that you will be wrong only about once in twenty times in a large number of trials.

L. NATURAL SELECTION

Understanding evolution requires a minimum knowledge of the mechanisms of population biology. Thus, before listing science requirements one by one, it is useful to summarize the biological and vocabulary knowledge that is involved in our descriptions and explanations.

Organisms create copies of themselves in great numbers. Their offspring are very like themselves because the information for the structure and function of an organism is incorporated in their DNA (RNA in some viruses, but functionally the same), rather like a recipe. The organism copies the recipe so that its offspring can proceed to reproduce as their own parents did and thus to continue the lineage. The enzyme that makes the DNA copies is a biological catalyst called a *replicase.* Occasionally it makes a mistake, which is termed a *mutation.* Mutations may change the ability of an individual offspring to survive and prosper in its environment. Most mutations will be harmful, but on rare occasions the change will be helpful. (Other mechanisms exist for producing changes in gene frequency, such as reassortment, recombination, conjugation, transformation, environment [radioactivity, carcinogens], or migration, but these factors need not concern us here.)

The reproductive machinery commonly makes far more offspring than can survive to produce the following generation. How will the choice among them be made? Those offspring with harmful mutations are likely to be eliminated without producing any offspring of their own. For example, if you are on a mountain at high altitude and can't get enough oxygen because of a mutant hemoglobin (or a mutant gene for any other important function, such as, e.g., feeding), you may not get off that mountain alive.

Moreover, if a different mutant offspring has a better oxygen-transport system than the one commonly available in the population, that mutant-carrying individual may be better able to survive and, therefore, on average, have more reproductively successful offspring than the current members of the population possessing the common, unmutated gene. Continued over numerous generations, the favorable mutant will increase in (relative) numbers and eventually completely replace the gene for the old enzyme or protein. This process is called *natural selection*. The favorable mutant is more fit, and one can measure its relative fitness as the ratio of the average number of offspring (progeny) the favorable mutant has to the average number for the unmutated gene. For example, if the unmutated members have 49 successful progeny for every 51 progeny for the favorable mutant, then the ratio is 51/49 = 1.04. The larger the ratio, provided it is greater than 1.0, the faster the new mutant spreads in the population and replaces the older variant. If the ratio is less than 1.0, the new mutation will be removed and that variant will become extinct.

It is important to understand that the creation of new mutations during the replication of an individual gene is quite rare. The probability that a particular nucleotide of human DNA is mutated (imperfectly duplicated) is only about one in a billion. That is an extraordinarily low error rate. But if you produce enough offspring (say, a billion sperm or eggs) or undergo sufficient numbers of germ-line replications, there is a high probability that any particular nucleotide position of the DNA sequence has a mutation in it somewhere in the population.

As lineages change over time, two subpopulations of a larger population may become increasingly different until, at some point, one subpopulation can no longer interbreed with members of a second subpopulation. The two populations are now different species. This process is called *speciation*.

But why should the two subpopulations get more distant from each other? Darwin first noted that (a) variation occurs among members of the same species or breeding population, (b) these variants are heritable, and (c) the more-fit variants leave more offspring for the next generation. Darwin then surmised that the combination of these facts was sufficient to account for the process of evolution—the noncyclic change of organisms over time, the essential features of what is called Darwinism. Neo-Darwinism is a modified Darwinism that incorporates new information about population dynamics and new genetic understanding (e.g., Mendel's) of material not available to Darwin. For our purposes here, they may be lumped together.

M. SOME INTRIGUING OBSERVATIONS TO CONSIDER

It will be useful to illustrate how the scientific process works with regard to questions related to those that Darwin raised. Dating has already been explained. Six intriguing questions obtained by observing the material world are listed in this section. They are followed in the next section by ten requirements that are needed for a scientific explanation of the observations. It will be shown how those general requirements are met by the theory of evolution. The point is to illustrate what science is and how science does what it does. It is hoped that the contrast with creationism sheds light on the distinction between science and nonscience, and gives a picture of how science is practiced.

1. It was seen above that fossils exist from at least 1600 M.Y.A. Actually, the oldest fossils known are from more than 3600 M.Y.A., with the age of the Earth being 4540 M.Y.A. Related fossils frequently form a continuum across several successive geological strata, but related fossils may also occur in strata that are not consecutive. These are frequently called *gaps*. In general, the older the strata, the fewer are the gaps and the larger is the length of those gaps. What are the fossils? How did they get distributed as they are?

2. The fossils show a continuing increase in their diversity (numbers of species) and complexity (number of organs and tissues) as time goes by. How did this diversity arise?

3. Many fossils of a given era often resemble the fossils of the preceding era, and the most recent fossils are clearly like the organisms seen around us today. Why should the fossils in upper geological layers (strata) be more similar to currently living organisms than to fossils from deeper geological layers?

4. The time span for these fossils (3600 M.Y.A.) is a very long time in terms of generations.

5. Moreover, on two separate occasions in 2009, Chinese workers reported finding a primate skull of about 65 million years of age, which roughly halved the previously known gap of 130 million years.

6. There are exceptions to the continual increase in diversity, in which it appears that a large fraction of the organisms disappeared suddenly. These are called *mass*

extinctions. Jack Sepkoski and David Raup have identified five major extinctions as being the most significant. These extinctions are as follows: End Cretaceous (65 M.Y.A.), End Triassic (205 M.Y.A.), End Permain (251 M.Y.A.), Late Devonian (375 M.Y.A.), and End Ordovician (450 M.Y.A.). Other extinctions have taken place, such as the extinction of megafauna in North America as recently as fifteen thousand years ago, but these are considered to be less significant. Some scientists, such as Bradley Cardinale, say that we are currently experiencing a sixth major extinction.

N. WHAT ARE THE GENERAL REQUIREMENTS FOR A SCIENTIFIC EXPLANATION OF THESE SIX OBSERVATIONS?

For an explanation of these observations to be scientific, it must meet, at a minimum, all of the ten following requirements:

1. There needs to be a material mechanism that shows *how mutations occur.* (What is the mechanism underlying the observed mutations?)

2. There needs to be a material mechanism that provides *continuity over time.*

3. There needs to be a material mechanism for creating *new biological forms* (plants, animals, fungi, bacteria, etc.).

4. There needs to be a material mechanism connecting requirements 2 and 3.

5. There needs to be a material mechanism by which some changes *become widespread.*

6. The explanation must be consistent with other known facts.

7. The explanation must lead to the *prediction of previously unknown events that can be tested* to see if the prediction, and thus the explanation, may be incorrect.

8. The explanation must have *generality.* This means that the explanation must explain many other events as well as the observed event.

9. The above mechanisms must *operate sufficiently quickly* that the time since the formation of the Earth is enough to account for the changes observed.

10. The explanation must be *a material or natural one,* because supernatural mechanisms are not allowed in science.

O. HOW ARE THESE REQUIREMENTS OF A SCIENTIFIC HYPOTHESIS MET?

We shall consider each of these ten requirements in turn, considering how they are explained scientifically.

1. *Mutations* occur when the replicase makes an error in duplicating the DNA recipe. The mechanism of mutation to yield new sets of similar genes, followed by natural selection among those mutations, is the general mechanism underlying all of the requirements in the previous list. The most general answer to the process is mutation followed by natural selection.

2. *Continuity* is required because offspring are so similar to their parents that there must be a way of faithfully transmitting information from one generation to the next. This is accomplished by

the handing down of copies of the DNA recipe from parent to offspring, from generation to generation. That also assures that the abilities of the offspring will be very much like the abilities of their successful parents.

3. *New species* arise because of the accumulation and spread of mutations of the DNA in the members of a population to the point where the members of one population can no longer interbreed with another. This process of dividing and divergence ensures that there will be new species arising with new abilities. Speciation events generally occur more frequently than extinction events, so there is a continual increase in the number and divergence of organisms.

4. The *connection* between points 2 and 3 is DNA. It provides a recipe for continuing the species on into the future while simultaneously dividing populations into species-forming groups.

5. Natural selection determines that some new variants *will increase in number* because their genes better adapt the organism to its environment and thus, on average, tend to produce more offspring that are fertile than other members of the population. Such processes eventually cause the complete replacement (fixation) of an old gene by a gene that better adapts the organism. This is a mechanism for the spread of more fit genes, which will increase the diversity of nature. Thus the great accuracy of DNA replication gives you continuity, while the mutation of DNA gives you new properties.

6. *Consistency:* Darwinian theory predicts that the order of divergence based on DNA analyses must correlate strongly with the

order of the fossils in geological strata. More precisely, it predicts that the more different two taxa are at the DNA level, the deeper in the geological strata their closest fossil ancestor will lie. It cannot be emphasized too strongly that *the correlation between fossil age and DNA divergence is very high.*

7. *New predictions:* In argument 8 below it is shown how one can get many sets of characters with which to test evolutionary theory. Evolutionary analytical techniques are increasingly capable of getting the apparently correct relationship, as shown by the same answer being obtained when using quite different data or organisms.

8. The evolutionary explanation has enormous *generality* because it applies to all forms of life. An example mentioned (in section I) is the observation that the bones and teeth of horse and cow are more similar to each other than either set of bones and teeth is to the bones and teeth of mice. That suggests that the horse and cow have a common ancestor, and that these are more closely related to one another than either is to the mouse. Said differently, the horse–cow ancestor is the most recent common ancestor of the three pairs of taxa.

But what does the generality requirement mean? It means that if one looks for other triples of characters for these three animals, they should suggest the same ancestral relation: that the horse and cow are the closest of the three possible pairings (figure 6). An example of a much different character is hemoglobin, a protein in blood that carries oxygen from the lungs to the other body tissues and is responsible for the red color of blood. What should one expect to see? If one believes in Darwinian evolution, one would predict that horse and cow hemoglobins should be more similar than either is to that of the mouse. Examination of the hemoglobin sequences shows that the prediction is indeed correct. Such a test on three animals has been done many, many times, enough that one can say that the ancestral relations of the three animals being considered are not at all disputed. And each new triple again is found to be consistent with the Darwinian prediction.

As usual, things are imperfect in biological tests. If one chooses many triples of characters for the gorilla, chimpanzee, and human triple, one gets human–chimp pairs most often, but one also gets a sizable number of chimpanzee–gorilla pairs plus a few human–gorilla pairs. The problem is that the species are all too closely related to each other for the amount of information available to separate them. The fault is not with Darwinism but with a lack of data—a lack of a sufficient number of varied characters. This can be resolved by increasing the number of characters. As the number of characters increases, the probability of the closest pair being the human–chimp pair increases toward 1.00.

9. The mutation rate can easily account for the new favorable mutants. It is difficult to be quantitative about this, but one can see in the literature the fixation of genes producing a large effect, meaning that, although the functional change seems very great, it may result from a mutation in a single gene. Moreover, it may be selected for and fixed very quickly. (*Fixing* means completing the substitution of an old gene with a new one.) Finally, some mutant genes may be fixed in a very small number of generations. Over several billion years of evolution, there are many more generations than would seem to be required for fixing even the slowest spread of mutations. The explanation is clearly materialist: no metaphysical processes or elements are being used.

10. A number of biological observations of interesting phenomena in the material world have been discussed here. A collection of general and specific requirements constituting an adequate explanation for those observations has been shown. As such, although omitting many relevant details, these observations and their associated explanations give a clear picture of the process by which evolutionists came to agree with Darwin. The data give overwhelming support for the Darwinian view—so much so that evolutionists are justified in calling it a theory in the most restrictive and scientific sense of the word.

CHAPTER FOUR

"Young-Earth" Creationism

A. BIBLICALLY BASED BELIEFS

What do some creationists believe that is different from the beliefs of others—particularly evolutionists and many other religious people, including the majority of Christians? Some of the following beliefs are no longer held by many creationists but do occasionally come up.

1. *The Bible Cannot Be Wrong*

Creationists believe that the Bible is the word of God Himself (who is perfect; Matt. 5:48: "Be ye therefore perfect, even as your Father which is in heaven") and thus it must be without possible error. This belief is obtained in part from 2 Timothy 3:16: "All scripture [the Bible] is given by inspiration of God." But the Bible also states, elsewhere, "That which I speak, I speak it not after the Lord" (2 Cor. 11:17; here, *after* means "from" or "according to" or "by command of.")

2. The Bible Must Be Read Literally

To understand the Bible, creationists say, the Bible must be read literally. An ongoing problem for centuries has been that of deciding how literally the present translations and versions of books of the Bible must be taken. The books of the Bible were written by many different people over many centuries, and were not combined into the collection that we have today until the fourth century A.D. Our present Bible is essentially a library of those books that were considered by theologians in the mid-fourth century A.D. to be divinely inspired. Other books that were once regarded with similar respect were considered untrustworthy by them and rejected. (Many of these still exist today, and make interesting reading.)

The common assumption is that the books included were all inspired by God ("All scripture is given by inspiration of God" [2 Tim. 3:16]), even though they were transcribed and written down by many different authors. A very relevant issue is how much of the wording in any one book is to be believed as having been divinely dictated, and how much of it reflects the writing and speaking style of the author at the time the book was written. And an equally significant related issue exists: how much of what we have today is a result of alterations in text during multiple recopyings (before the era of printing) and of translation from one language to another?

Most Christians and Jews today consider that the books of the Bible involve figures of speech and styles of expression in places—features that were never meant to be taken literally by the writers. When you are asked how your work went today, and you reply, "Ah, it was a real rat race!" no one imagines that you mean that you were miraculously changed into a rodent and spent the day in exhausting races with rodent opponents. To what extent should the many different writers of the books of the Bible be allowed the same latitude?

A few Christians reply to this question: "Never! Every word in the book should be, and must be, construed as the literal reality spoken by the Deity." This excessive literalism can give rise to some problems in logic.

The cornerstone of the creationist argument is that the events in the early chapters of Genesis are to be considered literally, word for word: true and unarguable. But a careful reading of the accounts of creation casts doubt on this assertion. The first two chapters of Genesis contain not one but two creation stories, and these differ in many respects. The first one, in Genesis 1:1–2:3, chronicles God's creation of the world in six days followed by a day of rest. Then, Genesis 2:4 recommences with a bridge passage—"*These are the generations of the heavens and of the earth when they were created, in the day that the Lord God made the earth and the heavens*"—which leads to a quite different account involving the creation of Eve from Adam's rib. The first version is sometimes called the "Elohim," or *E chronicle,* since God is designated by the Canaanite term *Elohim.* The second account is called the *J chronicle* since it uses the Judean name of God, *YHWH,* which is translated as either "Yahweh" or, more commonly, "Jehovah."

In the E chronicle, God is a distant, abstract creator who completes his work in six successive "days":

Day 1: Creation of light and darkness.

Day 2: Creation of the firmament and the waters above and below.

Day 3: Creation of dry land and seas. Creation of land plants of all descriptions.

Day 4:   Creation of the sun, moon, and stars.

Day 5:   Creation of sea creatures and bird life.

Day 6:   Creation of land animals, followed by male and
female human beings with dominion over other
life forms.

Day 7:   God's day of rest.

This version raises the question, mentioned earlier, of how day
and night could have existed before the sun existed, as well as
the even more contentious issue as to whether the "day" mentioned is a modern twenty-four-hour clock time interval, or a
figurative term meaning an "era" or some otherwise unspecified
period of time.

By contrast, in the J chronicle, God is a more interactive or,
one might say, "human" persona, who speaks directly to his cre-
ations and advises them. The order of events is somewhat different from that of the E chronicle:

Genesis 2:4   God creates the heavens and the earth.

2:5   All plant life is still dormant for lack of
              moisture.

2:7   God creates man (Adam).

2:8–9   God creates the Garden of Eden with irrigated
              and flourishing plant life.

2:15–17   God places Adam in charge of Eden, making
              only one tree forbidden.

2:18–20   God creates land animals and birds, and asks
              Adam to name each of them.

2:21–25   God creates woman (Eve) from one of Adam's
              ribs.

To summarize, the order of creation in the E account is (1) land plants, (2) birds, (3) land animals, and (4) male and female humans.

In J the order is different: (1) male human, (2) flourishing plant life, (3) land animals and birds, and (4) female human.

What is one to make of this contradiction?

The answer is that *the E and J chronicles are separate accounts, taken from different sources.* The E version has many parallels with a Babylonian creation myth that is known as the *Enûma Elis* from its opening words: "When on high..." The J chronicle, in contrast, is more closely related to an earlier Akkadian epic, the *Epic of Atrahasis* (named after the hero of the story). Both the *Enûma Elis* and *Atrahasis* are known today from cuneiform clay tablets that have been discovered at more than one excavation site in the Middle East. The author or authors of Genesis apparently began their work by combining the two most authoritative accounts of creation from the Middle Eastern society in which they lived. Successive scribes and translators naturally tried to smooth over the two accounts and meld them into a single chronicle, but if one goes back to the original language, the differences become more apparent: the use of *Elohim* versus *Jehovah* for the deity, for example. The fact that the break between the two accounts occurs in the middle of a chapter is irrelevant, since the chapter and verse divisions are not ancient; indeed, the forms we see today in the King James Bible were finalized by Archbishop Stephen Langton only around 1277 A.D.

In sum, if the creation story in Genesis is really the conflation of two separate stories, each stemming from an earlier Sumerian or Babylonian source, and if these two stories differ in many details, then *how can one logically assert that both of them are literally and simultaneously true, word for word?* The claim would be thrown out by any competent court of law.

3. How Long Is a Day as Used in the Bible?

> And God said Let there be light: and there was
> light. And God divided the light from the
> darkness. And God called the light Day, and the
> darkness he called Night. And the evening and
> the morning were the first day.
>
> Genesis 1:3–5

> And God made two great lights; the greater light
> to rule the day, and the lesser light to rule the
> night.... And the evening and the morning were
> the fourth day.
>
> Genesis 1:16–19

If you accept Genesis 1 literally, you must believe that day and night are not consequences of the relative motion of the sun and the earth, because day and night began occurring three days *prior* to the creation of the sun. What happened to cause and effect?

In English the word *day* has three meanings: (1) the time from sunrise to sunset (which is irrelevant here and will not be considered further); (2) the twenty-four-hour period between two successive high noons; and (3) a loosely defined but a considerably longer period of time—as, for example, "in Herod's day," meaning "in Herod's era."

As a critically important aside, the Hebrew words *ab* and *ben*, used in the genealogical "begat" section of Genesis, can mean "father" and "son," respectively, but they also are used properly in Hebrew to mean "ancestor" and "descendant." That could imply a considerable increase in Archbishop Ussher's (see next section) estimation of the time back to Creation Week; but this would not compensate for the much greater discrepancy between "young-Earth" creationist and scientific estimates of the age of the Earth.

Some creationists accept the third definition above as a way to resolve the difficulty between the 4.54 billion years that elapsed before humans appeared on this planet according to scientists and the six thousand years that have since elapsed according to creationists. (The value of six thousand years quoted here varies a bit depending on which creationist you read, and can be as high as ten thousand years.) The two estimates for the origin of the Earth—4.54 billion years versus ten thousand years—differ by a factor of nearly half a million. There are roughly as many years in the evolutionist's model *as there are minutes* in the creationist's picture.

Among the creationists there is a major division with respect to *when* the creation took place. Those that believe that it was about six thousand years ago are called "young-Earth" creationists or, sometimes, YECs. The others tend to allow a considerably older age for the Earth based upon the uncertainty of the meaning of the word *day* in the old Hebrew of Genesis 1. *Day*, they conclude, can sometimes be considered a metaphor for "era," as when we talk about how things were back in the Romans' day.

4. How Did the Creationists Arrive at Their Date of the Earth's Origin?

Archbishop James Ussher calculated the origin of the world as Sunday, October 23, 4004 B.C. This was based on the genealogy in Genesis, the various cycles of the sun and moon, and other sources such as historical writings and problems with the Julian calendar. Because Adam and Eve were born within a few days of the creation of the universe, the dates for both the world's and man's origin are the same. Clearly by creationist estimates the universe began 4,004 + 2,010 = 6,014 years ago as I write these lines.

Creationists oppose having a tree to represent life's evolution because it contradicts the assertion that God created each of the various kinds of life's organisms separately during Creation Week. Nevertheless, the tree structure is admissible for relationships within kinds such as humans, for without it one could not use the "begats" (Gen. 11) to determine the creationist's estimate of the age of the Earth.

5. Noah's Flood

Creationists claim that the flood endured by Noah really happened, and inundated the entire planet, including parts that were higher than the top of Mt. Ararat (Gen. 7:19: "and all high hills that were under the whole of heaven were covered"). They claim evidence for this by the discovery of fossils on high mountains. It does not prove the height of the flood, but it is consistent with that of high water. The ark, they say, touched down on Mt. Ararat on Sunday, May 5, 1491 B.C. The Biblical account is essentially a retelling of an earlier story from the *Epic of Gilgamesh*. The *Epic of Gilgamesh* is an important work of Babylonian literature whose origins go back to the third millennium B.C. As with the *Enûma Elis* and *Atrahasis*, it is known today from clay tablets with cuneiform writing that were excavated in the nineteenth and early twentieth centuries. It tells the tale of Utnapishtim, who was warned by the gods of a coming deluge and told to construct an ark into which Utnapishtim was to load "all my kith and kin, the beasts of the field, the creatures of the wild, and members of every skill and craft." When the rains came,

[t]he great gods of heaven and of hell wept, they covered their mouths. For six days and six nights the winds blew, torrent and tempest and flood overwhelmed the world, tempest and flood raged together like warring hosts. When the seventh day dawned the storm from the south subsided, the sea grew calm, the flood was stilled. I looked at the face of the world and there was silence; all mankind was turned to clay. The surface of the sea stretched as flat as a roof-top; I opened a hatch and the light fell on my face. Then I bowed low, I sat down and I wept; the tears streamed down my face, for on every side was the waste of water. I looked for land in vain, but fourteen leagues distant there appeared a mountain, and there the boat grounded. On the mountain of Nisir the boat held fast, she held fast and did not budge. One day she held, and a second day on the mountain of Nisir she held fast and did not budge. A third day, and a fourth day she held fast on the mountain and did not budge; a fifth day and a sixth day she held fast on the mountain. When the seventh day dawned I loosed a dove and let her go. She flew away, but finding no resting-place she returned. Then I loosed a swallow, and she flew away but finding no resting-place she returned. I loosed a raven; she saw that the waters had retreated, she ate, she flew around, she cawed, and she did not come back. Then I threw everything open to the four winds; I made a sacrifice and poured out a libation on the mountain top. Seven and again seven cauldrons I set up on their stands; I heaped up wood and cane and cedar and myrtle. When the gods smelled the sweet savour, they gathered like flies over the sacrifice.

The *Epic of Gilgamesh* was widely circulated in the Middle East during the second millennium B.C. In modern times it was first discovered in cuneiform clay tablets excavated from the royal library at Nineveh, in northern Iraq. Other versions of the story subsequently were found in excavations at Nippur, in southern Iraq; at the old capital of Akkad, in southern Turkey; at Sultantepe (also in Turkey); and at Meggido, in Palestine. The tale seems to have been a Babylonian "best seller" and was almost surely familiar to the author or authors of the book of Genesis. As with the creation story, the compiler of Genesis drew from earlier literary accounts to make his point. It is worth noting that the story of Utnapishtim and the flood is only a small part of the longer *Epic of Gilgamesh.* By the same token, the story of the creation is only a small component of the older *Epic of Atrahasis,* mentioned under heading A.2 above. Indeed, *Atrahasis* also contains a flood story similar to that of the *Epic of Gilgamesh.* Evidently, borrowing and reworking were standard practices in Babylonian literature— a tradition that was embraced by the Hebrews for their own purposes.

B. NONBIBLICALLY BASED RELIGIOUS BELIEFS
1. The Origin and Distribution of Fossils

According to "young-Earth" creationists, all living things were drowned except the pairs Noah took into his ark. The geological layers (strata) containing different fossils all stem from this flood. Creationists have suggested three processes to explain why different fossils occur in different geological layers. One process is that animals have different densities (relative weights) and those that were the most dense sank faster and thus occupy the lowest levels of fossilized layers today. The second process suggests that different animals move at different speeds, and those that moved the fastest got farther up the mountain before the rising floodwaters engulfed them. The third process suggests that organisms had different ecological zones (niches) and that, as the surface of the floodwaters rose to higher altitudes, the members of successive zones were drowned and fossilized.

It is interesting that creationists present three different explanations for Grand Canyon fossil orderings. These are material explanations that are eminently testable. What were the densities (specific gravities, actually) of the living organisms that are in the fossil beds? How fast could they move? How high above the ocean floor did they live? One may need to assume that these properties in today's sharks (and other fossilized organisms) are sufficiently similar to those of the Devonian sharks that the correlation of the strata fossils remains significant. In fact, quite different values between then and now can be tolerated provided there are not too many reversals of the rank orderings in the lists in which they appear. Creationists often complain that they are excluded from the major biological journals. Jonathan Wells has written that "defenders of Darwinian orthodoxy have managed to establish a near-monopoly over research grants, faculty appointments, and peer-reviewed journals in the United States" (2000, p. 236). These tests of hypotheses, *if properly carried out,* would surely get you into one of those journals whichever way the results came out.

2. The Second Law of Thermodynamics and the Evolution of Life

Entropy is a measure of chaos, of randomness, or of disorder in a system. The second law of thermodynamics says that, in all reactions *in a closed system,* entropy always increases. The creationists correctly note that living organisms have more organization and less entropy than the simple materials from which they are made.

Surely only God can produce works in violation of the second law. Thus God is proven to exist.

For the evolutionist, this is incorrect. Let us design a simple experiment. Put a flat-bottomed beaker in a stand above a table, fill it with water, and let it stand overnight. By morning the molecules of water will be going in every direction, behaving chaotically. Water molecules are too small to see, but you can see their effect on a few small (say, 0.5 mm) diameter pieces of cork. They will be seen to move in a chaotic fashion called *Brownian movement.* Now place a gas burner (a Bunsen or Fischer burner) under the beaker, and start heating the water in the beaker. (Do not boil!) In a short while you should see, from the dark and light patterns (called *schlieren*), that the water is now flowing in an organized loop. The warm, less dense water rises in one place while the cooler, more dense water sinks to the bottom, where it warms up, rises again, and repeats the cycle. The entropy of this organized flow is less than that of the previous randomly moving water. You have created an object, the water-filled beaker, *with less entropy than it had just before you started the heating.* You must be a God by the creationist reasoning in the previous paragraph.

What is wrong in all this is the failure to note the italicized phrase, "in a closed system," in the definition of entropy given two paragraphs back. If the reactants are enclosed in a perfect vessel such that neither matter nor energy can flow into or out of the vessel, that would be a closed system. The lower entropy of the organized water is bought at the price of much more entropy being created by the burning of the natural gas to provide heat from the burner. The entropy reduction in the beaker is possible precisely because it is an open system in which extra energy is consumed in heating the water while the gas is burned to carbon dioxide and water.

In a similar fashion, the lower entropy of living organisms is bought at a higher entropy price, as the organisms in an open system convert metabolites to carbon dioxide and water. The system exacts an even greater price when the sun's light energy is used to make chlorophyll and photosynthetic (plant) products. Finally there is yet more entropy created when herbivores (plant eaters) convert plants into sheep, goats, and cows. Thus, in an open system, a part of the system may have less entropy provided that other parts of the system gain even larger amounts of entropy.

Our planet itself is an open thermodynamic system. It is kept going by a constant inpouring of energy from the sun, just as the Bunsen burner poured energy into the beaker of water in the experiment above. Imagine that by some mysterious process the Earth was enclosed in an impenetrable shell that kept out all light, energy, and matter from the outside. Under those conditions—those of a closed thermodynamic system—life indeed could never have evolved, or even survived for long, had it once been created. But the Earth is not contained in a closed system, and life does not violate the principles of thermodynamics. (One prominent creationist advocate, who by his training should know better, regularly dismisses the above line of reasoning with a cavalier "Ah, here comes the closed-system argument again!")

3. Intelligent Design

Intelligent design is the idea that some things are so complex—the structure of some objects so carefully designed to perform an apparently useful function—that those things must have been designed by some intelligent designer, by some creator. The books of Phillip Johnson have the clearest exposition of this view, although the idea was well presented much earlier by William Paley (1802). The concept is that there are objects, such as life and the universe, that are so complex that no natural processes could possibly have created them. That raises the question, How can one know one is looking at intelligently designed objects?

Because of the many intelligently designed things around us (automobiles, skyscrapers, medicines, clothing, bridges, musical instruments, paper clips, fluorescent lights, zippers, books, fine arts, Norman arches, aqueducts, algebra, etc.), there must be millions of intelligent designers called humans. And the creationists need at least one other intelligent designer, some nonmaterial entity, to explain the origin and occurrence of living organisms, the planets, and the universe itself. No more than one supernatural being is necessary as long as he or she is powerful enough, knows how to create all these organisms, and can do so in six working days.

Such a being seems to be indistinguishable from God. Both this special designer and God are supernatural forces, in that they are above and beyond the naturally occurring forces that scientists study. Creationists advocating intelligent design often avoid the term *God* in their presentations. They appear to avoid any association between God and the intelligent designer in order that intelligent design not be seen as a religious doctrine, even though there are no discernible differences between the two. There is no apparent way by which one could distinguish between the works of God and those of supernatural intelligent designers. Until the creationists provide such a recipe, it seems reasonable to consider their intelligent designer to be God. Even in the presence of such a recipe, as long as the intelligent designer must have supernatural powers, it is not a materialist construction and thus is not scientific and therefore not appropriate to study in the scientific classroom as an explanation of observed complexity. Creationists present an argument that is highly analogical and thus cannot prove much of anything.

(As will be discussed later, either the traditional God is the intelligent designer, or there exist more than one deity with supernatural powers. See: "How might one choose between the two models: creationism vs. Intelligent Design" below.)

The intelligence that designed human beings seems rather less than perfect. Here is a partial list of human imperfections, many of which are from Olshansky, Carnes, and Butler 2001.

a. More check valves in leg veins would reduce the frequency and severity of varicose veins.

b. Thicker bones would reduce breakage.

c. More ribs would reduce hernias.

d. Tilting the torso forward would reduce the frequency of slipped discs.

e. Shorter stature would reduce fracture from demineralized bones (osteoporosis).

f. Removing the useless appendix would eliminate appendicitis, which sometimes kills people; hence the creation of the appendix is wasteful as well as harmful.

g. A larger pelvis would, when women give birth, better accommodate the large heads of human babies.

h. Male nipples are useless and thus a waste of resources.

i. Humans need larger jawbones or smaller/fewer teeth.

j. Rudimentary ear muscles are useless.

k. Hemorrhoids are useless.

l. Thicker spinal discs would reduce back pain.

m. The coccyx or tailbone is useless in humans.

n. Mutation rate is very high, requiring excessive genetic load (deaths).

o. Too many progeny unsupportable by the environment leads to too much selective death.

p. There are many harmful phenotype mutations born that are ill adapted to their environment and quickly die off without producing offspring.

q. Eyes that see in the ultraviolet would be very helpful.

r. Having the optic nerve pass in front of the retina creates the blind spot, which the octopus does not have because its optic nerve passes behind. In this respect the octopus's eye is more intelligently designed than that of humans.

s. An immune system reacting faster and with greater selectivity against invading pathogens would be very helpful.

t. Human males inefficiently make billions of sperm just to obtain one fertilized egg.

u. A bladder that better resisted its reduction in volume by an enlarging prostate would be wonderful.

v. An extra rib exists in approximately 8 percent of the population.

w. Many pseudo (nonfunctional) genes exist in the genome (the total DNA of an organism).

x. Cruelty and sadism are counterproductive and damaging behavior patterns.

y. Psychiatric dysfunctions are equally damaging.

z. The morality of humans needs considerable improvement.

Dembski (1999) has recently asserted that evolutionist complaints about poor designs seem to be that the designs are not optimized for the apparent function intended. He notes that *optimal* is not the same as *intelligent*. Thus although human beings have been designed, the human design is not everywhere optimal, and a more intelligent designer might have thought of numerous improvements.

The creationists contend that one cannot know the mind of God, nor that one can know for what use these structures may be intended at some future date. Thus the materialist belief that these features are suboptimal is merely an acknowledgement of our ignorance, and not indicative of a deficient design. The creationists are quite right about this, but the elegant, complicated structures then become of little use as support for intelligent design. If all assertions of apparently imperfect design can be defeated with the claim that the flaws only appear to be imperfections, that they stem from the evolutionist's (and the creationist's!) ignorance of God's divine purpose, then all logical arguments about intelligent design are doomed to failure. For if all criticisms are defeatable by invoking incomplete understanding, then intelligent design itself is not testable, and debates on intelligent design are pointless. Moreover, if one cannot know the purpose of the design, how can one know whether the structure is well designed? And if one cannot know whether something is well designed, how can the concept of intelligent design be used at all? And if the answer to asserted design flaws is that "one cannot know the mind of God," then intelligent design stands self-described as theological, and not the materialist construct that creationists assert intelligent design to be.

The creationist could also get around the problem of needing God to be perfect by letting some other, less than perfect entity (sometimes called the "demiurge") be the intelligent designer responsible for designing plants, animals, and other organisms. But if so, that creates a bigger dilemma for the creationist, for now God must be given up as the unique creator of humans—in direct contradiction to what is said in Genesis.

Creationists have also suggested that lists of numerous but modest imperfections such as the twenty-six listed above should be interpreted as the result of simultaneously optimizing multiple characters in which the deficit of any one character is the price paid for other characters so that the group is best overall. But that in turn implies that the omnipotent God is incapable of fully optimizing each. Moreover, it does nothing to solve the problem of not knowing the mind of God.

How might one choose between the two models: creationism versus intelligent design?

a. Assume that *at least one God is required to create any living thing* (otherwise evolution might create an organism).

b. Therefore, the existence of humans proves *there is at least one God*. (This conclusion is not required for the following argument, but it might appeal to some.)

c. Dembski (1999) has correctly noted that a design can be intelligent without being optimal. And certainly, as far as humans are concerned, a lot of intelligence would have been required to get humans to the current design stage. But the twenty-six imperfections listed above can

be dismissed only by twenty-six appeals to our ignorance of God's purposes, and the more often such
appeals are necessary, the less credible are the appeals.
Accordingly, *the design of humans is imperfect.*

d. If the design is imperfect, *the designer is imperfect* (for the
"tree is known by its fruit" [Matt. 12:33]).

e. If the imperfect designer exists, then *God need not be the
one responsible for these imperfections* (unless God approved
of the design).

f. But now there are two deities: God, and the imperfect
intelligent designer (see the assumption in *a*, above).

g. Therefore, *Christianity is polytheistic* and not, as the Bible
states, monotheistic.

h. However, if one assumes, alternatively, that God is the
intelligent designer, then *monotheistic Christianity is
restored.*

i. In this case, if *God is the intelligent designer,* then God is
Himself responsible for all those imperfections. (This
assumes that the strict creationist, faced with a choice
between accepting polytheism and the proposition that
there is no designer other than God, will choose the
latter.)

The group who call themselves Raëlians hold beliefs that
are similar to creationists except that, for Raëlians, the intelligent designer is not God but rather an extraterrestrial race of
peoplelike beings called Elohims, who created us as an experiment, carrying out that experiment on Earth so that their own
planet would not be contaminated by anything that might
evolve.

4. Irreducible Complexity

Irreducible complexity is the idea that something can be so complicated that it must have been designed. The general concept is for the purpose of arguing that organisms (or eyes) could not have arisen by any mechanism other than by an intelligent designer. Thus evolution is wrong and creationism is correct. (Note that this is another fallacy of the excluded middle, in that it is assumed that there are only two possible answers, evolution or creationism.)

Behe (2000) has illustrated that question with the idea of an irreducibly complex object, one of many parts that work together in such a way that the removal of any one part destroys the object's ability to perform its function. Behe's example is the mousetrap, which has five parts (platform, catch, hold bar, spring, and hammer), with the removal of any one part making the mousetrap nonfunctional. To prove that an object is irreducibly complex, every part's removal must necessarily destroy the function.

Consider a functional object with five parts: A, B, C, D, and E. Imagine assembling the object by some series of steps such as $A + B = \rightarrow AB$; $AB + D = \rightarrow ABD$; $ABD + E = \rightarrow ABDE$, and $ABDE + C = \rightarrow ABCDE$. This represents the case where C is only added in the final assembly step. Of course when we see the assembled object we don't know in which order the pieces were put together, so ABCDE, ABDEC, BADEC, and all other arrangements of the five letters describe the same final object. For simplicity, we will keep the letters within any object in alphabetical order.

In the absence of other information, any one of the five components could have been the final one to be added. That is, the next-to-final state of the object could be ABCD, ABCE, ABDE,

ACDE, or BCDE. But could any one, or even more than one, of these four-component forms be functional without the addition of the final piece? Or does the removal of any part of the object destroy its ability to function?

For Behe the interesting case is the one in which none of the five tetramers that lack one part is functional. That means that the object must be so complex that no intermediate can function properly. If no intermediate is functional, then natural selection could not have produced the object. But if none of the possible paths are suitable for natural selection, the object must have been designed, and hence creationism is the explanation.

The correctness of Behe's conclusion depends critically upon the assumption that all possible next-to-last intermediates have been tested. This is surely not so, at least to the extent that Behe's model is quite different from that of the evolutionist. Behe seems implicitly to have a model in which all five of the parts (A, B, C, D, and E) are first constructed independently in their present form, and only then are the parts fitted together to make the completed object.

The evolutionist's model is one where initially the cytoplasm (cellular juices) contains numerous polypeptides (short strings of amino acids) that are slowly changing. At some point two such polypeptides bind together, however weakly, and perform an action useful to the cell's survival. Mutations accumulate in the polypeptides, and the mutants' frequency increases if their weak binding is enhanced. Eventually a third polypeptide might bind, and the process is repeated as often as needed. The last (most recent) step in the process is very likely to be an amino acid replacement in one of the interacting polypeptides. The Behe model has a number of very large final steps, whereas the evolutionist's model needs only small steps. Behe correctly rejects his large-gap model, but that does not leave intelligent design the victor. This is another example of the logical fallacy of the excluded middle. The biological model wasn't tested.

A good illustration of the problem involved is hemoglobin, a carrier of oxygen in the blood. (See Dickerson and Geis 1983.) In bony fish and in all land vertebrate animals—amphibians, reptiles, birds, mammals—hemoglobin has four polypeptide chains or subunits: two so-called alpha chains with 141 amino acids each, and two beta chains with 146 amino acids each. The complete $\alpha_2\beta_2$ hemoglobin molecule is a small molecular machine. Each chain can bind one oxygen molecule to a heme group, which consists of a central iron atom embedded in a flat plate of carbon and nitrogen atoms known as a porphyrin ring.

Each heme group is packed in a pocket in the surface of one of the α or β subunits, and the subunits are packed against one another in a roughly spherical molecule. When, at the oxygen-rich lungs, oxygen molecules bind to one or two of the heme groups, the four subunits shift relative to one another in a way that makes the remaining two or three hemes particularly prone to binding oxygen as well. Conversely, when one of the subunits of a hemoglobin molecule drops its oxygen at the tissues, the molecule rearranges itself so as to unload the other three also. Because of this machinery of the hemoglobin molecule, the binding of oxygen tends to be an all-or-nothing proposition, making the four-chain hemoglobin molecule an especially efficient carrier.

So the hemoglobin of higher organisms is truly a molecular machine, an analog of Behe's mousetrap. But hemoglobin hasn't always been a four-chain machine. Lower organisms such as mollusks, marine worms, midge larvae, and even nitrogen-fixing plants such as soybeans and lupine have single-chain hemoglobins.

(It isn't clear why nitrogen-fixing plants need hemoglobin, but the best guess is that this is their way of scavenging and getting rid of any traces of oxygen that might interfere with the process of nitrogen fixation in their root nodules.) The individual hemoglobin chains in these organisms are all alike and do not aggregate into tetramers.

But there's yet another chapter to the story. Sea lampreys, one of the more primitive of marine animals, have only one kind of hemoglobin chain, but are halfway toward a molecular machine on their own. In the absence of oxygen, two identical subunits combine as a dimer. But the binding of oxygen alters the structure of the chains in a way similar to that found in tetrameric hemoglobin, and causes the two chains to fall away from one another. Conversely, when oxygenated lamprey hemoglobin dumps its oxygen molecules, it dimerizes once again.

So here we have three levels of mechanism for the hemoglobin machine:

1. Mollusk hemoglobin binds and loses oxygen on isolated chains.

2. Lamprey hemoglobin dimerizes without O_2 but the chains separate when O_2 binds.

3. Fish and land vertebrate hemoglobins have one arrangement of four chains when free of O_2, and a different arrangement when oxygenated, producing a useful all-or-nothing oxygen-binding behavior.

Comparisons of amino acid sequences among hemoglobin subunits in all of these organisms—soybeans and lupine, mollusks and marine worms, lampreys, bony fish, amphibians, reptiles, birds, and mammals—makes it obvious that they differ from one another by evolutionary changes in individual amino acids. The most primitive creatures had only one type of gene for hemoglobin. Lampreys have multiple repeated identical copies of their hemoglobin gene, which produce only one kind of hemoglobin chain. In fish and higher life forms, these multiple gene copies have accumulated mutations that lead to different protein chains: α and β in our example, but also variants of the α and β chains that are found in embryos and fetuses in the early stages of life.

It is almost as though someone had demonstrated to Behe that a mousetrap does indeed work best if it has a complete inventory of platform/catch/hold bar/spring/hammer but under certain conditions will also work nearly as well with only platform/catch/spring/hammer, or with platform/spring/hammer, or even with only platform and hammer! In spite of what Behe has claimed, what we just described has actually been proposed by John H. McDonald (2011). Some types of "incomplete" mousetraps can actually function. As McDonald remarks, "The mousetrap illustrates one of the fundamental flaws in the intelligent design argument: the fact that one person can't imagine something doesn't mean it is impossible, it may just mean that the person has a limited imagination." So Behe's irreducible complexity argument isn't really applicable to mousetraps. It also isn't relevant for hemoglobins, and for eyes, and for many other organs that creationists have invoked in an attempt to justify the need for a designer. Many other cases have been studied in which incompletely constructed systems do work, although sometimes not as efficiently. Behe's mousetrap analogy fails. A complex mechanism in a living organism does not have to be created, assembled, and installed in final form in order for it to do something useful for the host organism. But evolutionists can be gratified that Behe himself has confirmed their conclusion that the model he advocated is wrong.

Another interesting unit is the ribosome (the unit that makes proteins). The ribosome is one of the most complicated cellular complexes; it comprises (in the human intestinal bacterium known as *E. coli*) two subunits containing twenty-one proteins in the smaller subunit and thirty-four proteins in the larger subunit for a total of fifty-five different proteins. There are also several nucleic acids present in the complex. Curiously, one can remove all twenty-one small-unit proteins and still have a functional ribosome, although the reduced complex makes proteins much more slowly. Thus complexity is not easily estimated if the goal is to find a degree of complexity that permits you to know whether an organelle (any structure in a cell such as the nucleus, mitochondrion, or chloroplast) is, or would be, irreducibly complex. An apt analogy by Robert Sprackland (2006) is as follows:

> I turn the key, the engine starts. I do not know how the engine works but I accept that (today at least) it does. According to "irreducible complexity," the fact that I don't know how the engine works doesn't mean that someone else might know. Instead I am supposed to believe that God made the car. And therein lies the hubris [arrogance].

5. The Anthropic Principle

It has been noted that there are six constants in our universe that seem to be just right for life on this planet in that, if any one of the values were only a little different from their measured value, life as it is known could not exist here. The probability of getting all these just-right values is asserted to be so improbable that it must surely be the case that God made the values the way they are *just so humans could exist here*. But surely God could just as well have chosen to make humans optimal for whatever constants existed. These constants are as follows (Shermer 2006):

a. The total amount of matter in the universe.

b. The force with which the nuclei of molecules bind to each other.

c. The number of dimensions our universe possesses (must be three).

d. The ratio of the strengths of gravity and electromagnetism.

e. A number that, if smaller than 10^{-5}, produces a featureless universe (no stars, supernovas, comets, planets, etc.) and if greater than 10^{-5} would produce a universe of black holes.

f. A number that determines how fast the universe will expand or contract.

An alternative suggestion comes from those who think that the finding of values in close accord with current requirements for life as it is known is not persuasive evidence of a God responsible for it all. Namely, they suggest that we just don't know enough about limitations in our measuring instruments. An illustrative example follows.

There is a tribe of creatures who believe in the fish God, and every Frynight evening they hold services to show their gratitude for His delivery of fish for them to eat. The big moment in the service comes when the pooh-bah dips his sacred net into the sacred lake to retrieve a fish. It is set down in their sacred scriptures that the netted fish must be at least twenty-five centimeters long or their commune will die out. Following the verification of the length of the fish, the members of the religion come forth to an altar and partake of the blessed fish. Following this, the congregation goes about its business until the next weekly Frynight. The religion bases its faith in the observation that every time a fish is netted, it always measures at least twenty-five centimeters long. Only a God could intelligently design a fish that was always at least twenty-five centimeters long. And only a benevolent God would produce a fish exactly to the dimensions required by His tribe.

One day a different group of creatures arrived, and their pooh-bah dipped his net into the lake and pulled out a fish. It was only twenty centimeters long. The members of the first tribe interpreted this as a sign from the fish God that the newcomers were disfavored and so set upon them and killed them. As for the anthropic principle, the pooh-bah of the first tribe noticed that the mesh size of his net was larger than the mesh size of pooh-bah number two; but he never told anyone about it.

The argument is analogical, and it has already been observed that analogies cannot prove anything. It has been shown only that correlations between the values of the constants needed and those actually observed may be the result of unknown features in our instruments (e.g., net and mesh size). This does not prove that anthropic parameters have the correlations they do because of unknown properties of our instruments. But it does show that it is possible. For the rest of the story, there is nothing cogent for our creationist/evolutionist concerns except possibly some amusement.

C. MARK TWAIN AND RUPERT BROOKE ON INTELLIGENT DESIGN AND THE ANTHROPIC PRINCIPLE

Intelligent design is an insidious argument—one that some people have characterized as "creationism in a lab coat." If our world has been intelligently designed, then *by whom,* and *for whom?* The anthropic principle that we have already seen suggests that everything has been designed particularly for human beings: we are the crown of the creative process. Some early evolutionists such as Darwin's colleague and competitor Alfred Wallace also thought that man stood at the top of the evolutionary process, and that the whole thing had been for our benefit.

Mark Twain, whom we have seen before in chapter 2, poked fun at Wallace in 1903 in an essay entitled "Was the World Made for Man?" which is freely available on the Internet today. In it he suggested that at a time when the oyster was the grandest product of evolution to date, a reasonably intelligent oyster might have concluded that the goal of the evolutionary process had been to bring forth oysters: "and at last the first grand stage in the preparation of the world for man stands completed: the Oyster is done. An oyster has hardly any more reasoning power than a scientist has, and so it is reasonably certain that this one jumped to the conclusion that the nineteen-million years was a preparation for him; but that would be just like an oyster, which is the most conceited animal there is, except man."

The young English poet Rupert Brooke said much the same thing ten years later, in a poem that he wrote only two years before he died, a casualty of World War I.

Heaven

Fish (fly replete, in depth of June,
Dawdling away their wat'ry noon)
Ponder deep wisdom, dark and clear,
Each secret fishy hope or fear.

Fish say they have their Stream and Pond;
But is there anything Beyond?
This life cannot be All, they swear,
For how unpleasant, if it were!

One may not doubt that, somehow, Good
Shall come of Water and of Mud;
And sure the reverent eye must see
A Purpose in Liquidity.

We darkly know, by Faith we cry,
The future is not Wholly Dry.
Mud unto mud!—Death eddies near—
Not here the appointed End, not here!

But somewhere, beyond Space and Time,
Is wetter water, slimier slime!
And there (they trust) there swimmeth One
Who swam ere rivers were begun,

Immense of fishy form and mind,
Squamous, omnipotent, and kind;
And under that Almighty Fin,
The littlest fish may enter in.

Oh! never fly conceals a hook,
Fish say, in the Eternal Brook,
But more than mundane weeds are there,
And mud, celestial fair;

Fat caterpillars drift around,
And Paradisal grubs are found;
Unfading moths, immortal flies,
And the worm that never dies.

And in that heaven of all their wish,
There shall be no more land, say fish.

Those last two lines are so liturgical that one almost feels compelled to add "Amen." But the oysters were wrong, and the fish were wrong. What makes us think that *we* have any more claim to being the grand finale?

D. SPECIES ARE IMMUTABLE

Immutable here is relative, not absolute. The creationists oppose the concept that species can accumulate mutations and thereby be transformed into some new species because it contradicts the statement in Genesis that God created all living organisms during Creation Week: "And God made the beasts of the earth after his kind, and cattle after their kind and everything that creepeth upon the earth after his kind" (Gen. 2:24). The problem is that if a species arose by the gradual accumulation of mutations, then it would not have been among the creatures ("everything that creepeth upon the Earth") mentioned as having been created during Creation Week by God.

Thus the creationists do not have species but rather have *kinds*. Gish (1978, p. 35) has given a list of some of the creation kinds.

Invertebrates: protozoa, sponges, jellyfish, worms, snails, trilobites, lobsters, and bees

Vertebrates: fishes, amphibians, reptiles, birds, and mammals

Reptiles: turtles, crocodiles, dinosaurs, flying reptiles, and aquatic reptiles

Mammals: platypuses, opossums, bats, hedgehogs, rats, rabbits, dogs, cats, lemurs, monkeys, apes, and men

Apes: gibbons, orangutans, chimpanzees, and gorillas

Some creationists maintain that *microevolution* is acceptable (changing wolves to domestic dogs; both dog kinds) but *macroevolution* is unacceptable (changing Chihuahuas to cats; different kinds). Thus the various breeds of dogs from Chihuahuas to Saint Bernards belong to the dog kind. The mechanism that determines whether a mutation is an allowable microevolutionary change or a disallowable macroevolutionary change is not given. Weirdly, lobsters and bees are not only each a kind in themselves but also of the same kind (invertebrates). Protozoa and jellyfish are similarly related as a kind themselves but also of the invertebrate kind. It seems strange to think that a change from lobsters to bees (both invertebrate kinds) might be a microevolutionary event.

One might notice that the chimpanzees are a kind within apes, which are a kind within mammals, which in turn are a kind within vertebrates. There are therefore at least four levels of kinds in the kind hierarchy. There seem to be at least two different types of kinds, one of which Gish terms "basic." The word *basic* is not defined, so one cannot know in what way some kinds are basic and others not basic. None of this explains what material mechanisms would give rise to the kinds described in the Bible and preserve them over time from divergence into separate species.

Creationists Kurt Wise and Walter ReMine have developed a system called *baraminology*, which classifies animals according to groups called *created kinds*. Creationists who accept the classification systems such as baraminology reject the notion of universal common descent. Baraminology is considered a pseudoscience by evolutionary scientists, and is criticized for lacking peer-reviewed research or rigorous testing.

E. CUMULATIVE SMALL STEPS CANNOT ADD UP TO LARGE DIFFERENCES

There exists a presumption for which there is no material evidence—namely, that many small changes occurring in a species cannot add up to large differences. Michael Behe has likened this to crossing a hundred-foot canyon by means of jumping to numerous buttes no more than ten feet apart from one another. To quote Behe: "Many people have followed Darwin in proposing that huge changes can be broken down into plausible, small steps over great periods of time. Persuasive evidence to support that position, however, has not been forthcoming" (1996, p. 15). Most creationists do not dispute the idea that small change (microevolution) does occur. Small change can be observed in laboratory experiments and in the field. There is persuasive evidence that these small changes accumulate over time, as seen in common morphological structures, DNA sequences, similarities in embryo development, the fossil record, and the geographical distribution of species. Macroevolution is simply cumulative microevolution.

F. THERE IS TOO LITTLE TIME
TO EVOLVE SO MUCH

To support their belief that evolution does not occur, the creationists look for flaws in the evolutionist's argument. Here the argument is that you can't get there from here in the time available.

That, of course, is true if you believe that the Earth came into being only six thousand years ago. But if we think in terms of more than 3 billion years, it is hard to argue that there is not enough time, particularly since no good estimates exist of how much evolution could occur in a given unit of time. However, even single nucleotide substitutions have been known to produce major anatomical changes.

G. THE NET EFFECT OF ALL MUTATIONS
IS HARMFUL

Morris (1974) concludes that "the net effect of all mutations is harmful." From this one derives another reason that macroevolution cannot have occurred. But that ignores the fact that many laboratory experiments have been performed that give rise to mutations that confer an advantage to the possessor of that mutation. The creationists' assumption implies that only laboratory mutations can be advantageous and a source of useful mutations, whereas mutations in the wild, outside the laboratory, can only be harmful—a most unlikely situation. But in fact there is no evidence that all mutations are harmful, in either the laboratory or the wild. It would seem to make more sense if "harmful" in the first sentence of this paragraph was preceded by "often."

H. MANY RELATED FOSSILS ARE WITHOUT TRANSITIONAL FORMS: THE GAP PROBLEM

The fossil record is the subject of a great deal of argument between scientists and creationists. For creationists, the fossil record contains so many gaps or "missing links" (an outdated term) that any phylogeny (the evolutionary relatedness of organisms) based on such data is surely largely wrong. For example, Kurt Wise has written that "[t]he rarity of transitional species in the fossil record seems to fit the expectations of young-age creationism better than it fits the expectation of evolutionary theory" (2002, p. 125). But scientists are aware that the fossil record represents only a tiny fraction of the organisms that once existed. Many organisms (e.g., soft bodied) leave no fossils at all, and the environmental conditions for forming good fossils are rare. In addition, evolution does not follow the so-called chain-of-being model that progresses steadily from one organism to the next "higher" organism in the fashion of a ladder.

Evolution has a structure that is more like a bush. Two species can diverge from a common ancestor instead of developing from one into another in a continuous chain of intermediaries. Therefore, one should not commonly expect to find fossils with morphological features that are midway between any two selected species, nor should one think of the fossil record as being complete. Although paleontologists do not look for "missing links," they do occasionally find fossils that exhibit characters that are a mixture of known species, which help inform us as to evolutionary development. For example, the fossil *Archaeopteryx* is considered the oldest, most primitive bird yet found. However, *Archaeopteryx* has more in common with ancient theropod dinosaurs than with modern birds. Like the modern birds, *Archae-opteryx* has feathers and a wishbone. Yet, unlike today's birds, *Archaeopteryx* also has teeth and a long bony tail. One could examine other *Archaeopteryx* characteristics (e.g., clawed toes) and find various features that resemble either a modern bird or an ancient theropod dinosaur. *Archaeopteryx* is often used as a model of an organism that preserves ancestral features while exhibiting descendant novelties. These features give us insight into how birds descended from dinosaurs, as do recent fossil discoveries in China of feathered nonavian dinosaurs. It is certain that *Archae-opteryx* is at least a close relative of the direct ancestor of birds, and should not be ruled out as being the ancestor itself. Some modern bird features, such as hollow limb bones, had already evolved in theropod dinosaurs long before *Archaeopteryx*. Conflicting evidence needs to be weighed together under the principle of parsimony. It is not known how much evolutionary divergence had already occurred at the time *Archaeopteryx* lived (about 150 M.Y.A.).

The fossil record is incomplete, but new discoveries continue to be found. The April 6, 2006, issue of *Nature* (Daeschler, 2006) features a Devonian fossil that sheds light on the evolution of the tetrapod (four-legged vertebrate) body plan. The creationist's logical error regarding gaps is that of believing that the absence of a complete fossil record indicates that what has not been observed has never existed. But the absence of evidence is not evidence of absence. That logic would lead us to conclude that God does not exist if we decide that we can find no evidence of His existence. Evolutionists may not yet know all the details of how an eye evolved, but that does not mean that the eye didn't evolve.

What science has discovered about the eye is striking in its implications. What may be called a "master eye-building gene" has been found that controls the growth and development of eyes. In fruit flies (*Drosophila*), the gene is called the "eyeless gene." (When studied in 1915, it was associated with the absence of compound eyes.) The expression of this gene initiates the expression of several thousand other genes involved in eye development, and it can even cause an eye to develop at an abnormal location on the fly's body (on the wings, legs, or antennae). A homologue gene in mice (the "PAX6" gene), when induced in the fruit fly, causes additional fly eyes to grow on the fly's body at the site of the gene expression. It is startling that the PAX6 mouse gene should be able to cause development of a fly eye on the body of a fruit fly, despite 500 million years of separate evolution in mammals and insects. Genetic control mechanisms for eye building appear to be quite universal across the animal kingdom.

I. THE PALUXY EVENT

Dinosaur tracks have been observed in the Paluxy River bed in Texas. At one point it was claimed that a human footprint had been discovered alongside the dinosaur tracks. This is important because, if true, it means that people and dinosaurs coexisted. It would also demonstrate that there is a very big error in the current view of evolution. But these fossil tracks have been shown to be fraudulent, and creationists rarely mention them any longer. Nevertheless, a member of the Institute of Dendrology, in the Polish Academy of Sciences, recently cited these "footprints" as new evidence against evolution (Giertych 2006).

J. THE PILTDOWN AFFAIR

In 1911, Charles Dawson was given some bones found by workmen at an excavation site near Piltdown, in southern England. They were judged to be from a creature having intermediate characters between modern man and chimpanzee. By the knowledge of the time, this is what theory said they should be looking for, and the find was greeted with excitement. Gradually, over forty years, evolutionists became suspicious and finally proved that this fossil, too, was a fraud, including the presence of chemicals on the bones used to make the fossil look older. These chemicals are not found around Piltdown and so must have been added. Later on it was established that the fossil was actually a mixture of parts from a human, a chimpanzee, and a gorilla planted as an elaborate hoax.

It is not known who concocted these hoaxes or why, but evolutionists corrected the Piltdown error while also correcting the error of the Paluxy footprints—both of them being fine examples of the self-correcting character of science. Creationists have not yet disavowed their belief in the Paluxy event, although they have toned down their references to it.

K. THE RADIOACTIVE CLOCK IS ALL WRONG

The discrepancy between the age of the Earth as determined by creationists and scientists could be removed if the rate of decay of the radioactive isotopes were much higher at a previous time, making a young mineral look very old. The increase in rate of decay would have to occur for all radioactive isotopes simultaneously in order to provide the same amount of correction. No evidence has been obtained that would suggest that changes in rates of radioactive decay are at all possible. Indeed, such changes would appear to contradict the basic materialist assumption that the laws of nature do not change.

L. SELECTIVE CITATIONS QUOTATION

A number of evolutionists have been quoted out of context. The usual circumstance is to claim that the experimental results appear to support creationism when in fact they do not. Niles Eldredge, a paleontologist on the curatorial staff of the American Museum of Natural History, describes how he was getting little variation from his study organism, the trilobite *Phacops rana,* even though the ages of his fossils varied over 8 million years. This is not what was anticipated by the then-current thinking, and Eldredge's effort to understand why he was getting the "wrong" answers eventually led to the punctuated equilibrium proposal. This proposal delighted the creationists because it suggested that evolutionary explanations, and hence evolution, were wrong. For example, Phillip Johnson quotes Eldredge as writing that evolution "never seems to happen" (Johnson 1997). Johnson also hints that evolutionists falsify data when he quotes Eldredge writing that "the pressure for results, positive results, is enormous" (Johnson 1997, pp. 60–61). Clearly, Eldredge did not try to make his data fit the current orthodoxy, but instead proposed a new scientific explanation. John Pieret, in a 2004 Web post, explains Phillip Johnson's selective quotation of Eldredge:

> Eldredge was, of course, looking to obtain his Ph.D. and advance his career, Johnson pointed out, and it is hardly surprising that he

was feeling desperate to find out what was "wrong" about what he was doing; why he was not finding what he "should" under the best theory at the time. But pressure of this sort is hardly unique to scientists. Should we assume that, say, aspiring lawyers [Johnson's own profession], if faced with an inability to fit their knowledge within the questions asked on the bar exam, would take a "doubtful" course of action by cheating "if they expect to have successful careers"?

In fact, *despite* having found differences in the trilobites he [Eldredge] was studying, he did *not* try "to construe a doubtful fossil as an...evolutionary transitional" in an attempt to salvage his career. Nor did he ignore contrary data in the face of his admitted "underlying assumptions." Instead, he faced up to the data, presented his results and eventually participated in extending, instead of "propping up[,]" the prevalent view. While I prefer to believe that this was the result of Eldredge's personal honor and commitment to his profession, there could hardly have been a different outcome. Eldredge's conclusions and data would have been scrutinized closely, not only by the thesis examiners but, should he have tried to have had it published, as he was all but required to do, it would have been pored over by all the experts in the field. Trying to pass off a conclusion supporting the "received view" based on openly shared contrary data would have been the death of his career, not its salvation. (See Pieret 2004.)

M. EVOLUTION IS A TAUTOLOGY

Evolution is *not* a tautology. Consider the following syllogism:

Premise 1:   Natural selection is the survival of the fittest.

Premise 2:   Those that survive are the fittest.

Conclusion: Those that survive are those that survive (and therefore natural selection is a tautology).

The claim that evolution is a tautology is made by some creationists using the argument as presented above. That argument as written is indeed tautological. But it does not state what evolutionists believe and assert. For the evolutionist,

Premise 1:   Natural selection is the preferential survival and procreation of the more fit.

Premise 2:   The more fit are those better adapted to their environment.

Conclusion: Natural selection is the preferential survival and procreation of those better adapted to their environment.

Note that the second syllogism goes beyond the "those that survive, survive" type of circular reasoning. The meaningful concept of "adaptation" to the environment is introduced. We can surmise which characters may help organisms to adapt to the environment so as to survive and reproduce, and we can devise metrics for those characters. Thus, we have independent ways to assess which characters make an organism "more fit." We can make predictions of preferential rates of survival instead of being merely limited to waiting to observe which members survive. We can form falsifiable theories testing the chosen characters' supposed ability to affect the statistical outcome of preferential survival and procreation of organisms possessing those characters. For example, we may surmise that the length of a male bird's tail is a character that helps the bird find a mate and procreate. We can test the theory by randomly selecting male birds and trimming their tail feathers short, or pasting on additional feathers to lengthen the tail. Of course, what is described here is only a brief outline of an experiment. The experiment must be properly planned, with appropriate controls to ensure that we are effectively isolating the selected character. We can then determine whether, over time, a male bird with a longer tail has a statistical correlation with preferential procreation. Thus the creationist argument differs significantly from the argument that the evolutionists make, and the evolutionist argument is not a tautology. (See also the discussion of tautology in chapter 1, section D.10, on logical fallacies.)

N. EVOLUTION IS ATHEISTIC

The principles we put forward here assert clearly that science cannot address theological issues; they are in different (i.e., non-overlapping) areas of knowledge termed *magisteria* by Gould (1999). As noted by Pennock (2000), "Science is Godless in the same way as plumbing is Godless." Famous scientists such as Francis Crick, Richard Dawkins, and Peter Atkins are persuaded that there is no material evidence that God does exist. But when using inductive reasoning we occasionally can be led astray. The theory "all swans are white" is not proven by many viewings of white swans, and can be disproven by a single sighting of a rare but existing black swan. The "black swan" logical fallacy was used by John Stuart Mill in his discussion of falsification. There is also the problem of what David Hume called the "Principle of the Uniformity of Nature": supposing that events will continue in the future as they always have in the past. Our failure to find scientific evidence of God is not proof that God does not exist. It would be difficult indeed to develop a valid syllogism whose conclusion is "God does not exist." The most one can say is that we have not (yet) found scientific evidence that God does exist. Some creationists have committed the same logical fallacy in inductive reasoning by asserting that the absence of certain anticipated fossils is evidence that those fossils never existed. In both the case of the material evidence for God and the case of evidence of certain transitional fossils, we do not know whether no such data ever existed or whether not enough samples have been examined to date. The absence of evidence is not evidence of absence. We may see our theory disproven with the next black swan we encounter.

O. NON-INTELLIGENT DESIGN

Intelligent design is one of the few proposals that creationists have suggested that is actually materialistic, provided we change the definition of intelligent design to involve a new materialist selection mechanism. The evolutionary mechanism is based on selection by the environment. The new mechanism could be based on selection by a human agent (watches, cathedrals) or an unspecified agent (to make the eye of the octopus, or to make life). We would like a sizable number of characters that are different between an evolutionarily and an intelligently designed, created object. We could ask of any object, "What are the characteristics of the object? How many were evolutionary-like, and how many were intelligent design–like?" This would be a good start toward the goal of a quantitative estimate of the probability of the object being the result of intelligent design rather than evolution.

Unfortunately, the creationists have presented no characters whose states or patterns differ between evolutionary mechanisms and intelligent-design mechanisms. Thus we apparently cannot test the two mechanisms against each other. At the level of description at which the creationists have described intelligent design, it is not testable. By one of the rules that define science, a hypothesis must be testable. Thus intelligent design is not, as currently set forth, a scientific hypothesis. Coupled with the observations in section B.3 in this chapter, it must be concluded that the intelligent designer has no features that distinguish her or him from God. Nevertheless he or she has the creative powers of God, and is thus supernatural and therefore is (a) God.

There is one difference in pattern between intelligent design and evolution. Figure 7 shows a simple phylogeny for the vertebrates. Between any two adjacent interior nodes of the tree (the circles) are listed character changes (the squares) that arose during the period encompassed by the two adjacent nodes. It has the unusual feature that in all cases the new character occurs in all animals that trace their ancestry to the upper of the two circle nodes being considered. Moreover, all of the animals that have these new character states are members of a single clade (a method of classifying organisms into groups); that is, they are all more closely related to each other than to any other organism. The presence or absence of one of those character states immediately tells you whether the animal is a member of the upper or lower group. The reason for this is that organisms pass on their genes to their offspring, so that all descendents of the new mutant have the mutation, but no other animals do. (See chapter 3, section I.)

The situation for the human intelligent designer (say, an architect) is quite different. If one Norman architect invents an improved arch for the doorway, any members of the architectural community are free to adopt or not to adopt that arch in design-

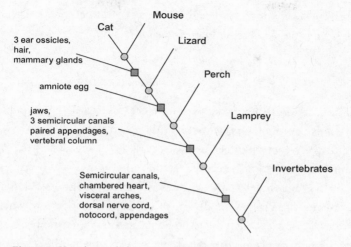

Figure 7. Vertebrate cladogram: a cladogram for five kinds of vertebrates. Each level of hierarchy (denoted by branch points) is defined by one or more similarities interpreted as evolutionary novelties.

ing their next building. Thus, given a large number of characters, each of which has at least two character states (electricity or not, indoor or outdoor plumbing, brick or wooden house, skylight or atrium, gas or electric heating, telephonic devices or not, etc.), each architect, using her or his own creative spirit, can and will mix and match, producing a series of buildings that do not have the distributional characteristics of biological organisms. With enough data points we should eventually be able to say whether biological organisms are significantly different from objects that are known to be designed.

Of course the creationists could, given the results of the test in the preceding paragraph, argue that God's design is different from human design. But that should mean that they can give us examples of humanly designed objects illustrating intelligent design as opposed to biological design.

P. TEACHING OTHER THEORIES

The creationists have sought to have creationism taught in public schools by influencing school boards to bring this about through textbook selection or getting laws passed that require the teaching of creationism. These efforts have been opposed by evolutionists. The legal response to those efforts of the creationists is to cite the First Amendment of the U.S. Constitution, which states: "Congress shall make no law respecting an establishment of religion, or prohibiting the free exercise thereof." This is more commonly called the *establishment clause* and is interpreted to require the separation of church and state. The laws so passed to teach creationism in science classes have all been declared unconstitutional because they violate the separation of church and state. This requires the legal system to find that creationism is a religious theory, not a scientific theory—and so it has, multiple times. (See "Kitzmiller v. Dover Area School District" 2005.)

The evolutionists have been less successful in opposing textbook choices. The current tactic is to have the books labeled with a "warning." ("This book contains material on evolution. Evolution is a theory, not a fact, regarding the origin of living things. This material should be approached with an open mind, studied carefully, and critically considered.") Of course, as we have already discussed, Darwinism is not a theory in the sense of a guess but rather a theory in the sense of having thousands of times been tested and having passed all the tests. Moreover, we have shown that biological evolution is a fact because the observations match the definition of evolution.

The matter of whether a work is scientific or not is an important issue. How many major works are scientific, and how would mandating the teaching of all of them affect teaching good biology? Examine the table below.

Theory	Status
Creationism	Unscientific even if it were proven true
Raëlianism	Unscientific even if it were proven true
Irreducible complexity	Unscientific even if it were proven true
Intelligent design	Unscientific even if it were proven true
Phlogiston	Scientific though proven false
Blending inheritance	Scientific though proven false
Orthogenesis	Scientific though proven false
Lamarckism	Scientific though proven false
Darwinism	Scientific even if it were proven false

Examples of ideas not from biology that, like creationism, are also not taught because they are wrong include: astrology, the ether, phlogiston, transmutation of metals, the four fluid humors of the body (blood, phlegm, choler, and bile), earth-air-fire-water as the basic elements of which all other material things are composed, most methods for predicting the future (tarot, tea leaves, crystal balls, Ouija, palm reading), and others. It makes for a bad educational method to spend much time teaching materials known to be scientifically wrong—except to use them as bad examples.

Evolutionists do not normally go around denying scholarly provisions for teaching creationism. Evolutionists are, however, interested in keeping unscientific theories and religions out of scientific classrooms. Creationism, if taught at all, should be taught in a comparative religions course that comprehensively covers the five religions with the most members (Christianity, 1,900 million; Islam, 1,100 million; Hinduism, 780 million; Buddhism, 320 million; Sikhism, 10 million; plus 1,200 million secularists ["Top 10 Organized Religions" p. B2]). But if more religions were wanted, the following is a partial listing of some others: Taoism, Dianetics, Druidism, Gnosticism, Jainism, Pantheism, Shamanism, Shintoism, and Zoroastrianism. How could one and why would one satisfy them all? Most evolutionists would support the offering of a comparative religion class in every high school curriculum. It would be enlightening and useful for students to see, in a setting of tolerance, the similarities and dissimilarities among the world's great religions.

Q. AN ATTEMPTED MATERIALIST TEST OF GOD'S EXISTENCE

It should not be assumed that there is no way of testing for God's existence. What we need is a result that differs depending upon whether or not God is doing something measurable. For example, many believers believe that praying to God for something might persuade God to grant the request. This is called *intercessory prayer.*

Assume that you have approximately twelve hundred patients who are going to have a coronary bypass—without, it is hoped, any complications (Benson 2006). We divide them into two groups, each of about six hundred patients.

Group 1 was told they might or might not have prayers said on their behalf, but all of them did. Group 2 was told they might or might not have prayers said on their behalf, and none of them did. For those for whom prayers were said, the prayers were said for fourteen consecutive days, beginning the night before the surgery.

The null hypothesis, the hypothesis being tested for, is that there is no prayer-answering God. The alternative hypothesis, for prayerful intervention, is that there will be significantly fewer patients with complications in Group 1 (the group that had been prayed for). If there are significantly fewer patients with complications, then we can reject the null hypothesis that there is no prayer-answering God.

In Benson's study, Group 1 (prayed for) had 315 complications out of a total of 604 patients (52 percent with complications), while Group 2 (not prayed for) had 304 complications out of a total of 597 patients (51 percent with complications). Although the percentage of complications was slightly higher in the "prayer" group, the numbers were not significantly different from each other, so we should not reject the hypothesis that there is no prayer-answering God. Nor did we prove that there is no prayer-answering God. We simply could not rule that possibility out. The test was inconclusive in this case. Benson tried to test whether or not there is a God that grants intercessory prayers. Had the "prayer" group shown substantially fewer complications, this could have been considered evidence suggesting the existence of a deity who intervened. But with both groups' results statistically identical, no conclusions can be drawn at all. There *might* be a God who decided for unknown reasons not to intervene in this particular situation. Or, it could be that God answers only one in a thousand prayers and our sample size was much too small to detect such a small bias in the statistics. However, Benson's study does suggest a method by which evidence might be gathered suggesting the existence or nonexistence of a God who intervenes when someone prays to Him.

R. GENETIC ALGORITHMS

There is clear laboratory evidence for experimental evolution in the lab. Moreover, this evolution can be commercially useful. An *algorithm* is a set of rules used to solve a particular problem. The *genetic algorithm* imitates the process of evolution by using a cyclical process in which each cycle leads to an improvement for some desired trait. Suppose that you have a piece of DNA that codes for an enzyme that performs, but very poorly, some necessary reaction vital for the cell. We now (cycle 2) make millions of copies of the DNA, but under conditions that replace one of its nucleotides with one of the other three nucleotides in 10 percent of those million copies. One expects that about 75 percent of the altered DNA sequences will code for an enzyme with a changed amino acid. We then test whether any of the new sequences have greater enzyme activity than the original DNA sequence. Choosing the most favorable new sequence (provided it is more active than the previous best sequence), we use it to repeat the process over again. Repeating the process (cycle 3) on the new sequence gives yet another sequence to use in the process. The number of cycles is arbitrary, with the overall result being an amino acid sequence encoding a much more active enzyme.

Not only has this process been used successfully many times; it has also been used in cases where no biological material was present, as when trying to optimize the fit of real data—say, economic data—to a polynomial equation with many variables. Thus a genetic algorithm (including natural selection) is a normal quantitative procedure of wide applicability and utility. There is every reason not to be surprised at its presence in evolutionary considerations.

Epilogue

The Literal Meaning of Genesis

And finally, let the last word be that of St. Augustine (A.D. 354–430) in his treatise *The Literal Meaning of Genesis* (which is highly recommended to all readers), as translated by J. H. Taylor:

> Usually, even a non-Christian knows something about the earth, the heavens, and the other elements of this world, about the motion and orbit of the stars and even their size and relative positions, about the predictable eclipses of the sun and moon, the cycles of the years and the seasons, about the kinds of animals, shrubs, stones and so forth; and this knowledge he holds to as being certain from reason and experience.
>
> Now it is a disgraceful and dangerous thing for an infidel to hear a Christian, while presumably giving the meaning of Holy Scripture, talking nonsense on these topics. We should take all means to prevent such an embarrassing situation, in which people show up vast ignorance in a Christian, and laugh it to scorn. The shame is not so much that an ignorant individual is derided, but that people outside the household of the Faith think that our sacred writers also held such opinions, and (to the great loss of those for whose

salvation we toil), the writers of our scripture are criticized and rejected as unlearned men.

If they find a Christian mistaken in a field which they themselves know well, and hear him maintaining his foolish opinions about the Scriptures, how then are they going to believe those Scriptures in matters concerning the resurrection of the dead, the hope of eternal life, and the kingdom of heaven? How indeed, when they think that their pages are full of falsehoods on facts which they themselves have learnt from experience and the light of reason?

Reckless and incompetent expounders of Holy Scripture bring untold trouble and sorrow on their wiser brethren when they are caught in one of their mischievous false opinions, and are taken to task by those who are not bound by the authority of our sacred books. For then, to defend their utterly foolish and obviously untrue statements they will try to call upon Holy Scripture for proof, and even recite from memory many passages which they think support their position, although they understand neither what they say nor the things about which they make assertion.

To this one can only add, "Amen!"

ACKNOWLEDGMENTS

In this book I hope to show how logic is involved in answering scientific questions about the world we live in. I want to demonstrate how biological evolution is a scientific study, even though it must inevitably be a work in progress with incomplete areas that are constantly being filled in by this logical process. And I want to show that evolutionary biology and reasonable religious beliefs are entirely compatible and not at all contradictory. In writing this presentation I have had help and encouragement from a great many colleagues, and it is appropriate to mention them here and express my appreciation.

I would like especially to thank Dr. Richard Dickerson, who over the past twelve months has worked tirelessly with me to strengthen and polish this manuscript and prepare it for publication. I am also grateful to my wife, Chung Cha, who from the very inception of this project worked hard to help me complete the manuscript.

Many of my colleagues in the Department of Ecology and Evolutionary Biology at the University of California, Irvine, have graciously taken time from their busy schedules to review early drafts of the manuscript. Drs. Michael Clegg and Rowland Davis reviewed the manuscript, gave me encouragement and showed great excitement about this project. Dr. Anthony Long helped me make the text more approachable to young minds by suggesting additional figures to aid in visualization. Dr. Debra Mauzy-Melitz was a great help in critiquing and constructing the figures. And Dr. Michael Rose commented, after reading the manuscript, "I am hopeful that your book will play a major cultural role in the near, if not immediate, future."

Among my U.C. Irvine colleagues I wish to particularly recognize and thank Dr. Francisco Ayala, who was kind enough to spend his valuable time suggesting methods to improve this book. To him I am very grateful.

I would also like to thank Dr. Glenn Branch, Deputy Director of the national Center for Science Education, who supplied comments and criticisms of early drafts. Dr. Al Bennet, Dean of Biological Science, U.C. Irvine, provided much appreciated financial assistance to help produce this work. Dr. Richard Siegel (emeritus, UCLA) read through early versions of the manuscript and encouraged me with his positive response.

And finally, my special thanks to Tom Dobrzeniecki, who has been my editorial assistant and research assistant for the manuscript. He has processed what must have seemed to him to be endless versions of each chapter. I deeply value his patience in going through the editing process.

GLOSSARY

α AND β CHAINS Names given to the two types of polypeptide chains in a hemoglobin molecule, having 141 and 146 amino acids, respectively.

ALGORITHM A set of rules describing how to solve a particular problem.

ALLELE When the two copies of a gene in a eukaryotic organism are distinguishable (for example by one copy having a mutation), the different copies are called alleles.

ANEUPLOID Having unequal numbers of the different chromosomes. Infants with Down's Syndrome have an extra chromosome 21. (See *haploid* and *diploid*.)

ANTHROPIC PRINCIPLE The idea that life as we know it could have arisen only if many of the constants of the universe were nearly exactly as they are, and thus the universe must have been created for us and that creator is God. This occurs because processes necessary for our continued existence would go either too slow or too fast if these constants were different. The conclusion is possible only if no alternative sets of constants would permit some different type of life. This is another example of the fallacy of the excluded middle (for which see the logic section).

BLACK HOLE An immense star that has collapsed into a very dense object whose gravity is so great that light cannot come out of it.

BLENDING INHERITANCE An early idea of how inheritance worked. It well explained cases where crossing a white flower with a red flower would give a pink flower. Unfortunately, if one crossed two pink flowers, one might get 25 percent white flowers, 25 percent red flowers, and 50 percent pink flowers— which was hard to explain under blending inheritance.

CAMBRIAN Time period dating from approximately 545 M.Y.A. to 495 M.Y.A.

CATALYST A substance that speeds up a chemical reaction.

CENANCESTOR The most recent common ancestor of a set of taxa.

CENOZOIC Time period dating from approximately 65 M.Y.A. to the present.

CHI-SQUARE (TEST) A statistical test to determine whether a theoretical frequency distribution sufficiently matches the observed frequency distribution.

CHLOROPLAST A subcellular structure in plants, responsible for capturing light energy and storing it as starch; life could not exist on Earth without this or some similar energy-capturing mechanism.

CLADISTICS Phylogenetic classification system in which a group of organisms share a more recent common ancestor with each other than they do with any organism not belonging to their group. Also called a *monophyletic group.*

CREATIONISM Defined here as the belief in the literal interpretation of Genesis, especially as it opposes Darwinian evolution on theological or other nonmaterialistic grounds.

CRETACEOUS Geological period from approximately 142 M.Y.A. to 65 M.Y.A.

CYANOBACTERIA Previously called *blue-green bacteria,* this is a phylum of bacteria that obtain their energy through photosynthesis.

CYTOCHROME C An iron-containing protein molecule made up of about 104 amino acids and found in the cells of every oxygen-breathing species.

CYTOPLASM The fluid that fills the cell and in which the organelles float.

DARWINISM A theory that provides a materialistic mechanism to explain the observed evolution seen in the fossil record. Its basic form accounts for evolution by the continuing creation of variation in populations and natural selection that favors some variants through giving them more offspring relative to other members of the population.

DEDUCTION Reasoning from the general to the specific.

DEISM The belief that God set the universe going but has not since intervened. (See also *theism.*)

DEVONIAN Time period dating from approximately 417 M.Y.A. to 354 M.Y.A.

DIMERIZE To combine two objects into one, as when two hemoglobin chains bind together in a molecule of lamprey hemoglobin.

DIPLOID Having two copies of each chromosome in the cell. Eukaryotes are diploid. (See also *haploid* and *aneuploid.*)

DNA Deoxyribonucleic acid; the biological material that encodes the instructions for making copies of a cell (the biological recipe for a cell).

DOVER A legal decision ruling that creationism is not science.

ENTROPY A measure of chaos, of randomness, of energy that cannot be used to perform useful work.

ENZYME A biological catalyst (generally a protein) capable of increasing the rate of some biochemical reaction.

EQUILIBRIUM, PUNCTUATED The idea that the rate of evolution is a mixture of rates, sometimes being a fast rate for short periods of time, punctuated by longer periods of time when the rate is very slow.

ESTHETICS The study of beauty and the fine arts.

ETHICS The study of right and wrong.

EUKARYOTES Those organisms that almost always have a nucleus, mitochondria, and two copies of their chromosomes, one copy from each parent. The mitochondrion was created when two different species of bacteria joined together to their mutual advantage. Such joinings are called *symbioses.*

EVOLUTION Noncyclic change over time, with application to cars, mountains, language, biological organisms, and so forth. "Noncyclic" is intended to exclude changes such as spring, summer, fall, winter, spring—changes that clearly do not reflect an evolutionary process but are cyclic.

FALSIFY To disprove a hypothesis.

FITNESS The average number of fertile offspring of a group divided by the number of offspring averaged over all members of the population; the greater that number (if greater than 1.00), the more rapidly the group gene spreads.

FIXATION The time at which a recently mutant gene finally replaces the last of the old genes.

GAMETE A haploid cell that is either a sperm (male) or an ovum (female; egg). Their union creates the fertilized cell, called a *zygote.*

GEIGER COUNTER An instrument for detecting and counting radioactive disintegrations.

GENE A nucleic acid that codes for some physiological or genetic function.

GENOME The total DNA of a single haploid cell of an organism.

GENOTYPE The genetic structure of an organism. (See also *phenotype.*)

GERM-LINE The serial set of reproductive cells making up the gonads (cells that produce gametes) and alternating with the gametes (sperm and ova) as the sperm fertilize the ova that produce a next generation of gonads.

GONADS The organs in which the reproductive cells are located

GROUP, MONOPHYLETIC See *clade.*

HABITAT The place where a group of organisms lives.

HAPLOID Having one copy of each chromosome in each cell. Bacteria are haploid. (See *diploid* and *aneuploid.*)

HEMOGLOBIN The protein molecule that is the oxygen carrier of the bloodstream, bringing oxygen from the lungs to tissues where it is needed. In almost all living animals it has four polypeptide or protein chains, each of 141 to 146 amino acids. In some primitive organisms it has only two chains, or even a single chain.

HERITABILITY The degree to which a complex, multicharacter trait (phenotype resulting from the effects of more than one gene on that trait; for example, height) appears in the progeny.

HOMOLOGOUS Having a common ancestor.

HYPOTHESIS, NULL The hypothesis to be tested. If the hypothesis is rejected, then you know that the hypothesis is not true. If the hypothesis is not rejected, the hypothesis may or may not be true. One cannot prove the null hypothesis. You can only reject or fail to reject the null hypothesis.

INDUCTION Reasoning from the specific to the general.

INTELLIGENT DESIGN The belief that some things (say, a cathedral, a temple, or a mosque) are not the result of chance processes but rather the result of conscious design. By analogy, the same logic permits one to conclude that life is similarly the result of conscious design. No designer other than God has been proposed. (See also *irreducible complexity*.)

IRREDUCIBLE COMPLEXITY Where the loss of any one component from a multicomponent complex destroys the function of that complex. The assumption is that if something is irreducibly complex, it must be the product of intelligent design. (See also *intelligent design*.)

KIND A creationist term similar to the evolutionist's *genus*. Dog is a kind, and it includes coyotes, hyenas, foxes and wolves; cats are a kind, but carnivore (cats plus dogs) is not a kind.

LAMARCKISM The belief that attributes of an organism that are modified after its birth can nevertheless be transmitted genetically to its offspring. For example, if one were to cut off the tail of generations of mice, eventually mice would be born that had short tails. This has been tried, and it failed.

Rudyard Kipling wrote a series of short "Just So Stories" for children. These delightful stories are very Lamarckian—as, for example, "How the Elephant Got Its Trunk." The trunk was created from a tug-of-war between a piglike animal and an alligator that had the other animal by its snout. As the tugging continued, the snout lengthened until it became the elephant's trunk.

MACROEVOLUTION The creationist idea that evolution cannot transform one organism into another if they are not of the same "kind." (See *microevolution*.)

MAGISTERIA Different (nonoverlapping) areas of knowledge.

MAMMALS A group of organisms, of which humans are members, that have hair covering their bodies and that suckle their young.

MATERIALISM The study of the physical world's matter and energy, and the interchanges of the two; also called *naturalism*.

METHODOLOGY, METAPHYSICAL Methods appropriate to the supernatural world.

METHODOLOGY, NATURALISTIC Methods appropriate to a materialistic world.

MICROEVOLUTION The creationist idea that evolutionary processes can only transform a member of one kind into another member of the same kind (as within the kind dog, where dogs include all the domestic breeds plus wolves, hyenas, coyotes, and foxes) but cannot transform one kind of organism into another (macroevolution) if they are not of the same kind (e.g., cats and dogs). It has the important consequence that Noah has to take only a pair from each kind, which reduces (but doesn't eliminate) the problem of how Noah got two members of all those organisms plus food for forty days aboard the ark, plus ways to get rid of all the excrement.

MITOCHONDRION A subcellular structure in eukaryotes that generates energy for the cells.

MOLECULAR CLOCK Refers to the estimate of the number of molecular changes over time in the hopes of determining a constant rate of change.

MUTATION Any change in the composition, order, or location of some of the DNA nucleotides of an organism.

MORPHOLOGY The study of an organism's form and structure.

M.Y.A Abbreviation for "million years ago."

NATURALISM Frequently used term that is synonymous with *materialism,* which see.

NICHE The location, space, and environment in which a coexisting group of organisms live.

NUCLEUS A structure of the eukaryotic cell that contains the genes of an organism.

OBJECTIVISM The assumption that there is an external reality.

OMNIPOTENT All-powerful; capable of performing all possible tasks.

OMNISCIENT Knowing all things.

OMPHALOS (Greek: "umbilical cord," "navel"). The hypothesis, proposed by Philip Henry Gosse in 1857, that Adam had a navel. Gosse further hypothesized that for the Earth to work, it must have been established with the *appearance* of age to function correctly. Thus the world has the appearance of age even though it is young.

ORGANELLE Any subcellular structure or compartment (e.g., nucleus, mitochondrion, chloroplast, etc.) floating in the cell's cytoplasm.

ORTHOGENESIS The idea that the evolution of organisms was directed by internal factors independent of environmental factors.

OVUM (pl., OVA) A female gamete, the male gamete being a sperm. The union of the male and female gametes, accomplished by the entrance of the sperm into the ovum, results in the fertilization of the ovum. The fertilized cell is called a *zygote.*

PARSIMONIOUS Requiring the fewest possible events to explain a phenomenon.

PHENOTYPE An observable structure or process of an organism; examples might include vertebrae and photosynthesis; see also *genotype.*

PHYLOGENY A representation, usually in the form of a tree, of the ancestral relationships of a group of organisms; similar to a genealogy.

PLATE TECTONICS A branch of geology that is devoted to studying the folding and breaking apart of the Earth's crust.

POLYPEPTIDE CHAINS Linear chains of linked amino acids that fold into a compact unit to form a protein molecule.

PRAYER, INTERCESSORY A prayer asking God to intervene on someone's or something's behalf.

PROGENY The offspring of a parent or parents.

PROKARYOTES Bacteria having only one copy of their chromosome, no nucleus, and no mitochondria. (See *eukaryotes.*)

PROTEIN A sequence of amino acids, usually capable of performing some biological function

PUNCTUATED EQUILIBRIA The concept that a group of organisms may remain relatively unchanged for long periods of time and then have a short-term burst of rapid evolutionary change; usually contrasted with a continuous gradual change.

RAËLIAN MOVEMENT Holds that life on Earth is the consequence of being seeded by a race of extraterrestrials, the Elohim. It is not otherwise much different from creationist beliefs. (*Elohim* means "God" in Hebrew but is said by Raëlians to mean "those who came from the sky.")

REDUCTIONISM The breaking of a larger problem into smaller problems.

RELIGIONS, ABRAHAMIC The group of the three major Middle Eastern religions: Islam, Judaism, and Christianity; so called because Abraham was a patriarch of all three religions.

REPLICASE A biological protein catalyst that makes copies of nucleic acids.

RIBOSOME A protein/RNA complex that functions in the synthesis of polypeptides.

SCIENCE The search for nonmiraculous explanations (rules) of the material world and the processes observed there. It cannot address metaphysical issues.

SILURIAN The third period of the Paleozoic era, from approximately 443 M.Y.A. to 417 M.Y.A.

SOCIAL DARWINISM The belief that the theory of Darwin applies to situations such as the affairs of human society without regard to ethics or morality.

SPECIATION A process wherein a single population of organisms diverges along two different lines of descent until the two lineages are so different that the they no longer interbreed with each other. The two lineages are then treated as separate species, and speciation has occurred. It is an obvious mechanism for increasing the diversity of organic life.

SPECIES A group of organisms that interbreed with others of their group but not with any members outside their group.

SPERM A male gamete.

STARLIGHT PROBLEM If the universe is only six thousand years old, how do we get light from stars a billion light years away? (See *omphalos*.)

SUBUNIT (PROTEIN) In those proteins that have more than one chain, one of the folded polypeptide chains that build a protein molecule. The mammalian hemoglobin molecule, for example, has two α and two β subunits.

SYLLOGISM A form of deductive reasoning consisting of a major premise, a minor premise, and a conclusion.

TAUTOLOGY A statement that must be true. For example, "Either dogs and cats are of the same kind or they are not." Definitions are sometimes circular because the subject and the predicate are the same, as in "A rose is a rose." Such statements are looked upon unfavorably, because the conclusion is already in the premise, and so that nothing new is produced. See chapter 1 for a more complete explanation.

TAXON (pl., TAXA) A species or, more generally, the smallest organic unit in a study.

TELEOLOGY A synonym for *purpose, goal,* or *end*. In biology it includes the assumption that organisms might act purposefully to change their genetic makeup. This assumption is generally rejected, although humans are beginning to recognize that they themselves might be able to do this.

TETRAMER An object with four similar components.

TETRAPOD A vertebrate with four appendages (arms, legs, wings, fins, etc.).

THEISM The belief that God does intervene in this world. (See *deism*.)

THEODICY The problem of evil. If God is omniscient and omnipotent, why is there any evil in the world?

THEOLOGY The study of God or gods and religious truths.

THEORY (OF EVOLUTION) 1. A well-supported explanation that is uniquely consistent with many thousands of observations. 2. A guess (colloquial usage).

TRUTH TABLE Analysis of a logic function by listing all the possible "truth" values ("TRUE" or "FALSE") that the function can attain. For example, if the operators p and q are considered with respect to the logical operation "AND," the table would list all possible combinations of truth values for p and q, together with the result of combining them with respect to the logical operator. Truth tables are used in Boolean algebra, Boolean functions, and propositional calculus.

UNIFORMITARIANISM The assumption that the laws of nature do not change.

VALID Said of a syllogism whose two premises, if true, correctly yield the conclusion (Tymoczko and Henle 2000).

ZYGOTE A fertilized cell. (See *ovum*.)

ANNOTATED REFERENCE LIST

The author's annotations follow selected entries.

Alters, Brian J., and Sandra M. Alters. 2001. *Defending Evolution: A Guide to the Creation/Evolution Controversy.* Boston: Jones and Bartlett.

Behe, M.J. 1996. *Darwin's Black Box: The Biochemical Challenge to Evolution.* New York: Free Press.

————. 2000. "Philosophical Objections to Intelligent Design: Response to Critics." www.discovery.org/a/445. Accessed July 12, 2011.

Behe, M.J., W.A. Dembski, and S.C. Meyer, eds. 1999. *Science and Evidence for Design in the Universe.* San Francisco: Ignatius Press. Introduced the concept of irreducible complexity.

Benson, Herbert, et al. 2006. "Study of the Therapeutic Effects of Intercessory Prayer (STEP) in Cardiac Bypass Patients: A Multicenter Randomized Trial of Uncertainty and Certainty of Receiving Intercessory Prayer." *American Heart Journal* 151, no. 4 (April 2006): 934–942.

Bronowski, J. *Science and Human Values.* 1965. New York: Harper and Row. A critique of logical positivism.

Copi, Irving M. *Introduction to Logic.* 1982. New York: MacMillan. A good introductory logic book.

Cracraft, Joel. 1983. "Systematics, Comparative Biology, and the Case against Creationism." In *Scientists Confront Creationism,* ed. Laurie R. Godfrey (New York: W.W. Norton & Company), p. 171.

Daeschler, E.B., N.H. Shubin, and F.A. Jenkins Jr. 2006. "A Devonian Tetrapod-like Fish and the Evolution of the Tetrapod Body Plan." *Nature* 440 (April 6, 2006): 757–763.

Darwin, Charles. 1859. *On the Origin of Species by Means of Natural Selection; or, The Preservation of Favoured Races in the Struggle for Life.* London: John Murray. The exposition of evolution that started the furor.

Dawkins, Richard. 1986. *The Blind Watchmaker: Why the Evidence of Evolution Reveals a Universe without Design.* New York: W.W. Norton & Company. Careful explanation of how chance events can give rise to the appearance of design in biological structures.

Dembski, William. 1999. *Intelligent Design: The Bridge between Science and Theology.* Downers Grove, IL: InterVarsity Press.

———. 2009. *The End of Christianity.* Nashville: Broadman & Holman.

Dennett, Daniel. 1995. *Darwin's Dangerous Idea: Evolution and the Meanings of Life.* New York: Simon and Schuster (pp. 314, 318).

Dickerson, Richard E. 1990. "Letter to a Creationist." *Science Teacher* 37:48–53.

———. "The Game of Science." 1992. *Journal of Molecular Evolution* 34:277–279. Also published as "The Game of Science: Reflections after Arguing with Some Rather Overwrought People," *Perspectives on Science & Christian Faith* 44:137–138.

Dickerson, Richard E., and I. Geis. 1983. *Hemoglobin: Structure, Function, Evolution and Pathology.* Menlo Park, CA: Benjamin/Cummings. A graduate-level monograph.

Dixon, Dougal, Barry Cox, R.L.G. Savage, and Brian Gardiner. 1988. *Illustrated Encyclopedia of Dinosaurs and Prehistoric Animals.* New York: Macmillan. Excellent reconstructed fossil drawings of the range of vertebrates, fish to mammals, with an emphasis on dinosaurs.

Einstein, Albert. 1954. *Ideas and Opinions.* New York: Crown.

Futuyma, Douglas J. 1998. *Evolutionary Biology.* 3rd ed. Sunderland, MA: Sinauer Associates. Standard evolutionary text for university undergraduate biology majors.

Giertych, Maciej. 2006. Comments in "Correspondence." *Nature* 444 (November 16, 2006).

Gish, Duane. 1972. *Evolution? The Fossils Say No.* San Diego: Creation-Life. Reprinted 1978.

Gould, Stephen Jay. 1999. *Rocks of Ages.* New York: Ballantine Books.

Gross, Paul R. 2004. "Patience and Absurdity: How to Deal with Intelligent Design Creationism," review of *Why Intelligent Design Fails: A Scientific Critique of the New Creationism*, by Mark Young and Tanner Edis (Editors). *eSkeptic: The Newsletter of the Skeptics Society*, no. 40 (October 29, 2004).

Hardin, Garrett. 1980. "'Scientific Creationism'—Marketing Deception as Truth." *The Dial* 1, no. 1 (September).

Hutton, J. "Theory of the Earth." 1788. *Transactions of the Royal Society of Edinburgh* 1:209–305.

Johnson, Phillip. 1997. *Defeating Darwinism by Opening Minds.* Downers Grove, IL: InterVarsity Press. Johnson is the author of several books on, and a leading spokesman for, intelligent design.

"Kitzmiller v. Dover Area School District." 2005. TalkOrigins Archive, December 31. www.talkorigins.org/faqs/dover/kitzmiller_v_dover_decision.html. Accessed April 27, 2011.

Larsen, Edward J. 2004. *Evolution.* New York: Random House. Basically an evolution presentation, but it addresses creationist issues where appropriate.

Layman, Charles Stephen. 2005. *The Power of Logic.* New York: McGraw-Hill. A good introduction to logic.

Lyell, C. 1877. *Principles of Geology.* 11th ed. 2 vols. New York: Appleton. The initial work suggesting gradualism and uniformitarianism.

McDonald, John H. 2011. "A Reducibly Complex Mousetrap." http://udel.edu/~mcdonald/mousetrap.html. Accessed July 11, 2011.

Miller, Kenneth R. 1999. *Finding Darwin's God.* New York: Harper Collins. A scientist who believes in evolution seeks ways to reconcile the differences between evolution and creationism.

Moore, John A. 2002. *The Case of Evolution and Creationism.* Berkeley, CA: University of California Press. A readable presentation of the evolutionist's beliefs.

Morris, Henry. 1974. *Scientific Creationism.* 2nd ed. Green Forest, AR: Master Books.

National Center for Science Education. n.d. "Ten Significant Court Decisions Regarding Evolution/Creationism." http://ncse.com/webfm_send/60. Accessed April 27, 2011.

Newman, Robert C., and John L. Wiester. 2000. *What's Darwin Got to Do with It?* Madison, WI: InterVarsity Press. A cartoon version of the material, often with largely rhetorical presentations, with bad logic to boot. No index. See p. 65 for an automotive fossil story designed to ridicule scientists. It uses a bad analogy.

Olshansky, S.J., B.A. Carnes, and R.N. Butler. 2001. "If Humans Were Built to Last," *Scientific American* 284, no. 3: 50–55. Consideration of the many imperfections in the human body that an intelligent designer would not have made.

O'Neil, Dennis. 2010. "Chronometric Techniques-Part I." Last modified April 8. http://anthro.palomar.edu/time/time_4.htm.

Overman, Dean L. 1997. *A Case against Accident and Self-Organization.* Lanham, MD: Rowman and Littlewood. Has a section at the beginning about logic. The most scientific of the creationist books.

Paley, William. 1802. *Natural Theology; or, Evidences of the Existence and Attributes of the Deity.* London: J. Faulder. One of the earliest and best expositions of intelligent design.

Panda's Thumb. 2011. "Evolution of Direct Development in Echinoderms." http://pandasthumb.org/. Accessed April 27, 2011.

Pennock, R.T. 2000. *Tower of Babel.* Cambridge, MA: MIT Press.

———, ed. 2001. *Intelligent Design Creationism and Its Critics.* Cambridge, MA: MIT Press.

Pieret, John [catshark, pseud.]. 2004. "Another Dishonest Creationist Quote." TalkOrigins Archive, February 16. http://www.talkorigins.org/origins/postmonth/feb04.html. Details of a misquoting by Phillip Johnson (in his *Defeating Darwinism by Opening Minds* [Downers Grove, IL: InterVarsity Press, 1997]) of Niles Eldredge's account of his graduate school days working on trilobites.

Pigliucci, Massimo. 2002. *Denying Evolution.* Sunderland, MA: Sinauer Associates.

Plantinga, Alvin. 2001. "Evolution, Neutrality, and Antecedent Probability: A Reply to McMullin and Van Till." In *Intelligent Design Creationism and Its Critics: Philosophical, Theological, and Scientific Perspectives,* ed. Robert T. Pennock (Cambridge, MA: MIT Press), pp. 197–236. A theist responding to criticism of an earlier paper; the theology gets heavy.

Popper, Karl. 1959. *The Logic of Scientific Discovery.* London: Hutchinson.

Provine, William. 1988. "Scientists, face it! Science and religion are incompatible." *The Scientist,* September 5, p. 10. A prominent biologist-historian claims that the theory of biological evolution disproves the existence of God.

Rose, Michael R. 1998. *Darwin's Spectre: Evolutionary Biology in the Modern World.* Princeton, NJ: Princeton University Press.

Rowland, Robert. 1968. "The Evolution of the MG." *Nature* 217 (20 January).

Sanders, N.K., trans. *The Epic of Gilgamesh.* New York: Penguin Books, 1972.

Shermer, Michael. 2006. *Why Darwin Matters: The Case against Intelligent Design.* New York: Owl Books.

Spenser, Herbert. 1867. *Principles of Biology.* Pt. 3, chap. 12.

Sprackland, Robert. 2006. "Darwin's Theory Is a Fact." *Seattle Post-Intelligencer,* January 19. A discussion of creationism in the classroom.

Taylor, J.H., trans. 1982. *St. Augustine: The Literal Meaning of Genesis.* Vol. 1. Ancient Christian Writers, no. 41. New York: Paulist Press (pp. 42–43).

"Top 10 Organized Religions in the World." *Christian Science Monitor,* August 4, 1998, p. B2.

Tymoczko, Tom, and Jim Henle. 2000. *Sweet Reason: A Field Guide to Modern Logic.* New York: Springer Verlag.

Vail, Tom, ed. 2003. *Grand Canyon: A Different View.* Green Forest, AR: Master Books. Photographs of the Grand Canyon with creationist ideas. Was sold at Grand Canyon National Park; apparently it still is, but is not to be found among the scientific books. Master Books is the publishing arm of the Institute for Creation Research.

Watts, Alan. 1970. *Does It Matter?: Essays on Man's Relation to Materiality.* New York: Pantheon Books.

Weller, Tom. 1985. *Science Made Stupid.* Boston: Houghton Mifflin. Wonderful satire of evolution.

Wells, Jonathan. 2000. *Icons of Evolution: Science or Myth?*. Washington, DC: Regnery. A creationist attempt to show why several evolutionary concepts like homology, Darwin's finches, peppered moths, Haeckel's embryos, and the Miller-Urey experiment are wrong.

Whitcomb, John C., Jr., and Henry M. Morris. 1961. *The Genesis Flood*. Philadelphia: Presbyterian and Reformed Publishing Co. A major source of creationist views on the Noachian flood.

Wise, Kurt P. 2002. *Faith, Form, and Time: What the Bible Teaches and Science Confirms about Creation and the Age of the Universe.* Nashville, TN: Broadman & Holman.

INDEX

absence, evidence of, 132, 138–39
absurdity, 17
Adam/Eve, 39–40, 100, 101–2
ad hominem argument, 20–21
ad ignorantiam, 21
aesthetics. *See* esthetics
agnosticism, 42
Akkad, 106
algorithms, genetic, 146–47
alleles, 55, 56, 56 *fig. 2*
altruism, 35
amino acids: genetic algorithms,
  146–47; in hemoglobin, 119–21;
  racemization, 72–73; sequences
  of, in protein, 13–14, 61–65, 80–86,
  82 *fig. 5*, 86 *fig. 6*. *See also* DNA
  sequences; mutations, genetic
analogical reasoning, 9, 35, 112, 124
Anderson, Alan Ross, 26–27
anthropic principle, 122–27
Archaeopteryx, 131–32
arguments, analogical reasoning
  in, 9
assumptions, 47

atheism, 21–22, 42, 138–39
Atkins, Peter, 138
atomic theory, 11
Augustine, Saint, 149–50
authority, 19, 30–31, 39–40
automobile, evolution of the,
  49–50

baraminology, 129
begging the question (logical
  fallacy), 9–10
Behe, Michael J., 20, 117, 118–20,
  121–22, 129
Belnap, Nuel D., 26–27
Benson, Herbert, 144–46
Berra, Tim, 20
"Berra's Blunder," 20
bias, 24, 26
Bible: as creationist authority,
  30–31, 75, 76; creationist view of,
  3–4, 98–107; day length in, 103–4;
  as infallible, 98; King James
  version, 102; literal interpretations of, 99–102; as moral

Text:	10.75/15 Janson
Display:	Janson MT Pro
Compositor:	Westchester Book Group
Indexer:	Kevin Millham
Printer and binder:	IBT Global